MERRIMACK

The Resilient River

—— An Illustrated Profile of the ——
Most Historic River in New England

DYKE HENDRICKSON

America Through Time is an imprint of Fonthill Media LLC
www.through-time.com
office@through-time.com

Published by Arcadia Publishing by arrangement with Fonthill Media LLC
For all general information, please contact Arcadia Publishing:
Telephone: 843-853-2070
Fax: 843-853-0044
E-mail: sales@arcadiapublishing.com
For customer service and orders:
Toll-Free 1-888-313-2665

www.arcadiapublishing.com

First published 2021

Copyright © Dyke Hendrickson 2021

ISBN 978-1-63499-317-3

Typeset in 10pt on 12pt Sabon
Printed and bound in England

CONTENTS

FOREWORD

This book is part history and part call to action. The Merrimack River is one of the great assets on the North Shore of Massachusetts, but several years ago, it was named as one of the most vulnerable waterways in the country. Sometimes, it appears to be getting dirtier, not cleaner.

So *Merrimack, the Resilient River* has emerged as an examination of a valuable resource that is in trouble. It is not facing disaster. Visitors use kayaks, canoes, motorboats, and sailboats to enjoy the Merrimack. People fish. Some swim, though they should not do so after heavy rain. Thousands each day walk along the waterway or view it from the road. Close to half a million people get their drinking water from it.

Yet now is the time to be aware that the river needs care. Steps must be taken so it remains the beloved resource it has been for centuries. Numerous organizations are seeking improvement, including the Merrimack River Watershed Council and the Merrimack River District Commission. Hundreds of individuals are doing what they can, including removing trash and providing financial support to activist organizations. The health of this historic river is too important to ignore as it once was.

The Merrimack River has numerous credentials that qualify it as one of the "most historic" waterways in the nation. These include the following: the birth of the Coast Guard (Newburyport, Mass., 1790); the home of the industrial revolution (hydro-powered mills in Lowell, Mass., 1820s); the first planned industrial city (Lawrence, Mass., 1847); the discovery of a technology to produce clean drinking water (Lawrence, 1893); and the site of one of the first victories in organized labor (Lawrence, 1912). In addition to being a river of historical dimensions, it is also one of the most resilient rivers in the country.

For two centuries, it was seriously polluted. Due to the aggressive and filthy textile industries, this waterway became a hazard. Its waters often turned orange or green depending on the dye being used at a given mill; those who used it for drinking water often got sick.

Newburyport Harbor is a scenic site where the Merrimack River meets the Atlantic Ocean. (*Ethan Cohen drone photo*)

Fishing boats can still be seen in Newburyport Harbor, but there are fewer than ever. (*Dyke Hendrickson photo*)

For many decades, a community's wastewater would be sent directly into the river. That would include effluent from toilets and industrial waste from factories. Until the early 1970s, many communities possessed only the most rudimentary of the sewage-treatment plants. Considering the river was discovered by Champlain in 1605, with the region settled in 1635, this river has a long history of fighting the manmade elements.

The name Merrimack is believed to have been adopted by early European settlers from Merruasquamack, a name meaning "swift water place," which was given by Native American tribes for the portion of the river between Manchester, New Hampshire, and Lowell, Massachusetts.

Despite several centuries of industrial exploitation, the Merrimack has survived. Newburyport, building on its revitalized brick downtown and its location at the intersection of the Merrimack and Atlantic, is thriving. Salisbury is a haven for campers and fried-dough consumers alike; Amesbury is enhancing its location on the river each year. Haverhill, Lowell, and Manchester, N.H., have made remarkable commercial and residential comebacks. Nashua and Concord, N.H., are prosperous. Lawrence, too, has launched a civic drive to improve the downtown area, upgrade housing, and enhance education. Scores of smaller communities are also thriving.

Miles of farmland and forest along the river remain close to pristine. Marinas near the Atlantic are full of boaters using the river and the ocean. Some of the most beautiful private homes in Massachusetts are on the Merrimack, including dozens in West Newbury.

For all its checkered past, the Merrimack remains a key natural asset to much of Massachusetts and New Hampshire. Many U.S. rivers have gone through difficult times, of course. Captains of industry felt sending pollutants into the river was a tradeoff because the companies were providing an economic engine with many jobs. So, for years, it was highly polluted.

Yet the Merrimack has not only survived, but it has prospered. Most communities along the river are thriving. The river itself is providing recreational opportunities for many thousands. Contemplating the admirable Merrimack, it is troubling to realize that most Americans did not think of advocating for the environment until the 1960s.

Rachel Carson's *Silent Spring* rang a bell in the early '60s, and the Cuyahoga River (Ohio) fire in 1969 was an outrageous wake-up call to force political action to halt destructive polluting. The resulting Clean Water Act of 1972 kicked off a mandated, federally funded approach to cleaner water.

That was about fifty years ago and it was essential. Due to federal funding and supervision, the Merrimack is one American river that has been resurrected. Those in the dual-state Merrimack Valley are encouraged, but they realize that there is still work to be done. In 2016, the Merrimack was named one of the ten most vulnerable rivers in the country by a national environmental organization. After heavy rain, it can be polluted by discharges from some sewage-treatment plants.

Users must be wary. In 2020, some who boat, fish, and hike the Merrimack have seen effluent in the water after rainstorms. Dogs who romp in the river have acquired sores. One study focusing on Lawrence suggested that visits to the emergency rooms of that community increase after unsupervised youngsters swim in the river. Epidemiologist Jyotsna Jagai of the University of Illinois at Chicago School of Public Health and lead author of a 2015 study noted:

The river flows quietly past aging mills in Lawrence. (*Dyke Hendrickson photo*)

Existing infrastructure is already stretched beyond its ability to manage severe precipitation, and with climate change, extreme rainfall events are becoming more frequent, and so are combined sewer overflows.

These overflows can have serious health impacts on communities if untreated water carrying viruses and bacteria contaminate drinking waters.[1]

Yet recreation is not the only concern. Close to half a million residents get their drinking water from the Merrimack. It is scientifically treated, of course, but a cleaner river would make everyone a little more secure.

In addition to minimizing CSOs, advocates hope they can limit the entry of chemicals released by some companies near the river, and they want to slow the residential and commercial development that threatens wetlands, streams, and tributaries. Also, some municipal pipes in the larger cities still discharge debris into the waterway.

Yet looking back over a half-century, conditions are much improved, and you have to take the small victories where you find them. As a result of the Clean Water Act of 1972, the Merrimack has made a significant comeback.

Also, amendments to the Clean Water Act in 1987 may have saved rivers such as the Merrimack. President Ronald Reagan wanted to abolish federal grants for sewage treatment. Yet proponents, led by Sen. George Mitchell of Maine, were able to compromise with the president. As a result, low-interest loans are available for the construction of sewage-treatment projects. That legislation, too, was a victory.

This book examines some of the history of this waterway. It is not all-inclusive. It is a slice of what has happened over the years. Also, it captures some of the sentiments professed by those working to keep the river clean.

One goal of this tome is to stress that the political drive for clean water did not start until the '70s. The "environmental movement" is young compared to the age of the river. Political advocacy seeking a clean river must continue.

Another goal of the book is to project an appreciation of the river itself. Many North Shore residents drive across the waterway every day. Some walk along its banks. Others appreciate it from the windows of homes, restaurants, or public parks.

Yet many North Shore residents are not conscious of its vulnerability. *Merrimack, the Resilient River* reflects an appreciation of a natural resource that has enhanced the lives of North Americans for hundreds of years.

The Merrimack starts in Franklin, N.H., parts of which look like they did centuries ago. (*Dyke Hendrickson photo*)

In the summer of 2019, close to a dozen kayaks traveled from Franklin, N.H., to Newburyport. Here, state Rep. Jim Kelcourse and state Sen. Diana DiZoglio team up in a tandem. (*Dan Gravoac photo*)

Acknowledgments

The author wants to acknowledge the Bowlen Foundation, headed by lawyer James Connolly, for making grants available so this book could be written. The Bowlen Foundation has supported the Merrimack trilogy: *Merrimack, the Resilient River* in 2021; *New England Coast Guard Stories: Remarkable Mariners* in 2020; and *Nautical Newburyport: A History of Captains, Clipper Ships and the Coast Guard* in 2017.

He would also like to thank photographers Dan Graovac and Bryan Eaton for contributing excellent photos; to Newburyport librarian Sharon Spieldenner for help with computer issues; and to John Macone and Matt Thorne of the Merrimack River Watershed Council for their contributions. If a photo does not have a photo credit, it was likely taken by the author.

He would also like to thank the elected officials mentioned here for granting interviews and explaining issues. He would also like to acknowlede Sen. George Mitchell, a gracious friend of the family, for his work on behalf of clean water and air.

1

Now is the Time to Improve the River

Supporters of a clean Merrimack River gathered in the late summer of 2019 and kayaked the 117-mile length of the waterway to draw attention to the vulnerable condition of the historic river. Almost two dozen high-energy activists made the trip—or parts of it. They called themselves the "Voyagers." This group included local and state officials, educators, environmental leaders, chamber of commerce types, and a journalist.

The four-day journey generated significant media coverage. Their message was "the Merrimack is a natural resource that must be saved." The Voyagers declared it could be restored; they said it was a resilient river.

Response to the journey generated a variety of reactions among those who followed the news and online coverage. Some were intrigued by the well-intentioned trip. Others who saw news reports wondered why the trip was necessary. The latter attitude likely reflects sentiments on the North Shore. Some are vitally concerned about the health of the Merrimack; others barely take notice.

The Merrimack River begins in central New Hampshire and winds 117 miles southeast to the Atlantic Ocean in Newburyport, Massachusetts. The Merrimack River watershed is the fourth-largest watershed in New England, encompassing 5,010 sq. miles. It includes all or parts of scores of communities with almost 2.6 million people within its reach, according to the Environmental Protection Agency.

The quality of water in Merrimack has been compromised over the years. About two centuries ago, capitalists began building mills on its banks. Utilizing technology that included the invention of the cotton gin (1794) and the effective harnessing of hydropower (*c.* 1820), the river became the center of an industrial revolution that produced more cloth than any area in the country.

Yet the price of creating clothing included significant pollution of the river. Both liquid and solid debris were dumped directly into the waterway. Parts of the Merrimack sometimes took on the hue of whatever cloth was being dyed that day.

Newburyport's history goes back several centuries, as does its bridges.

Still, the Merrimack continues to be a beloved resource on the North Shore of Massachusetts and the southern portion of New Hampshire. It is a fast-moving river that scientists say has self-cleaning properties. One source said it moves 2 million gallons per minute. Remarkably, the river provides drinking water to close to 500,000 residents, and dozens of small hydro-power outposts still churn daily to provide energy.

It is a popular venue for recreation, including fishing, boating, camping, and bird-watching. Swimming is permitted in some areas. Miles of pathways permit pedestrians to walk along the river. The Merrimack is a focal point that unites those in many communities north of Boston.

Over the centuries, it has served as a source for commercial good. The harbor at Newburyport once was the shipbuilding center of the colonies. Communities on the lower Merrimack entered into international trade in a major way in the eighteenth and nineteenth centuries. It was said that some young men who had gone to sea for years-long journeys had been to the West Indies or parts of Asia before they ever visited Boston.

Riverside communities hosted hundreds of fishing boats. Some fished in the river itself, bringing in sturgeon, shad, salmon, and striped bass. Others sailed into the Atlantic, coming back with cod, tuna, halibut, and flounder. The lobster industry developed. Fish provided fresh food and income.

Depending on the decade one considers, the Merrimack has hosted thousands of boaters, fishermen, swimmers, kayakers, sailors, and pedestrians. Many appreciate the river for its changing beauty and consistent majesty. Taking a walk or parking a car along the

Merrimack is an attraction in itself. Indeed, at a seminar in early 2020, one observer declared that one of its values was as an "asset for mental health."

Yet in 2020, many who valued the river were ready to develop action plans to clean the Merrimack. City, state, and federal managers said it was becoming increasingly polluted by combined sewage overflows (CSOs). In an era of climate change, heavy rains in the spring and early summer send extra run-off to sewage treatment plants along the river. When the plants exceed capacity, managers release the entire contents. As a result, raw sewage enters the river as well as run-off.

The increase in thc CSOs is an alarming development because the river serves residents in life-sustaining ways. More than 500,000 North Shore residents get their (treated) drinking water from this river, and the Merrimack still provides the hydro-energy that creates electric power for thousands more.

Also, close to 2,000 pleasure boats are tied up or launched from Newburyport to Haverhill. This form of high-end recreation requires clean water. In Newburyport, the boat owners comprise a "second city" because hundreds of vacationers stay on their vessels throughout the summer.

The state's numerous boat launches enable thousands more to enjoy this natural resource. The boat launch at Cashman Park in Newburyport is the busiest state ramp in Massachusetts, and hundreds of restaurants and stores are dependent on consumers enjoying the river. Thousands come to Newburyport each year for Yankee Homecoming and fireworks over the Merrimack.

The year 2020 marks close to two centuries of mills operating on the river. Perhaps that is another reason why many residents of the Merrimack River Valley have a vision of a cleaner waterway. Parts of the waterway, once pristine and wild, had become polluted from 200 years of dumping from factories, sewage-treatment plants, and careless residents

Fishing off the banks of the Merrimack near Plum Island in Newburyport has always been a popular sport. (*Bryan Eaton photo*)

using it as a garbage lagoon. Bicycles, tires, and even whole automobiles have been pulled from its bottom. Needles discarded by drug users in some communities are particularly unsavory evidence of misuse.

Of course, observers have known about this pollution for a long time. Henry David Thoreau was among the first critics of mills that fouled the water. In 1839, he wrote a book titled *A Week on the Concord and Merrimack Rivers*, which was published in 1849.

The tome lauded river life that was found on the Concord River, but he was critical of the Merrimack. He wrote that the mill system was ruining the Merrimack by putting tons of waste into the water. It was fouling nature. Also, dams and canals were hindering the movement of fish. He wrote, "Perchance, after a few thousand years, if the fishes will be patient, and pass their summer elsewhere—nature will have leveled the Lowell factories, and the river will run clean again."

In modern times, Newburyport Mayor Donna Holaday has said at public gatherings, "After several days of rain, the river at our end can look brown—not blue." Newburyport itself has spent millions on upgrading its sewage-treatment system.

Yet somehow, this river has prevailed. The Merrimack survived during the factory era of the nineteenth and early twentieth century when the largest textile mills in the world were dumping debris straight into the water. It prevailed in the 1950s and '60s before federal funds and use regulations arrived. In 1972, the Clean Water Act was passed, and large federal grants helped pay for sewage treatment plans. The Merrimack started to get cleaner.

Henry David Thoreau, way back in 1839, warned that mills, and their attendant dams and canals, would ruin the fish runs.

By 2020, new threats had been recognized. Increased residential and commercial development—especially in parts of southern New Hampshire—are leading to the loss of natural filters so that run-off with grease and gasoline finds its way into the river. Also, some factories are sending toxins such as mercury into the Merrimack.

The regional branch of the EPA is aware of such misdeeds and issues permits for discharge, but an increasing number of North Shore residents are pushing for tighter enforcement.

Most activists feel the river is vital to the region, and not just for visual beauty. They are seeking federal funding to upgrade treatment plants. They want to separate rain run-off from effluent in riverside sewage-treatment plants. Some have suggested creating taxing districts that would generate a small sum from every household and business in the Merrimack Valley. Environmentalists say Portland, Maine, has been able to clean up Casco Bay by invoking such a levy. It is called a "rain tax."

In Newburyport, municipal officials are attempting to develop a system that would warn boaters and fishermen when there has been a CSO discharge upstream. Planners say that an app on a smartphone or perhaps flags at marinas and launch sites will alert recipients that foul water is on the way.

By 2020—about two centuries after the birth of the mill system in Lowell—a quiet activist movement was generating momentum for changes that would help the river. An awareness that the river is worth saving spread like a soft mist throughout the Merrimack Valley. Indeed, in one conference call during that year, more than 100 officials, environmentalists, and business professionals joined the conversation on determining what steps must be taken to clean the river and keep it that way.

One observation that the Voyagers made was that the river is still beautiful, and that nature prevails. Birds line the shores; ripples still punctuate the river. There are acres of forest adjacent to the waterway. The marsh at the eastern end, the largest in New England, remains a vibrant refuge for wildlife. In some parts of the river, swimmers still stroke in the Merrimack.

Parts of waterfronts are as grand as they were years ago, but the river requires constant attention and political support to keep it going.

Eagles have returned to the Merrimack, and in recent winters, the fishing has been good. (*Bryan Eaton photo*)

The river offers many scenic views, like this one in Methuen. (*Dyke Hendrickson photo*)

2

MERRIMACK RIVER:
THE THREAT

Anecdotal evidence suggests that the Merrimack River has serious problems, and state and federal officials often take samples of the river for testing, but what exactly is the threat?

American Rivers, a non-profit environmental organization based in Washington, D.C., says that the river is threatened by development, inadequate sewage-treatment plants, and the leakage of dangerous chemicals. This statement might serve as a starting point for those who want specific examples of the challenges that the river faces.

In 2016, American Rivers named the Merrimack as one of the ten most vulnerable rivers in the country. The remainder of this chapter is the assessment by American Rivers when it explained why the Merrimack was included on this unflattering list:

The birthplace of American industry, drinking water source for over a half million people, and home to Eastern brook trout and other fish and wildlife, the Merrimack River is one of New England's treasures. However, its forests are disappearing, cut down to make way for developments, roads, and parking lots.

Unless the EPA acts now to protect sensitive lands and implement green infrastructure solutions, the river and its communities will be choked by increasingly polluted run-off.

The health of the 117-mile Merrimack River and its 11 tributaries is of great importance to 2 million residents and an abundance of fish and wildlife species.

Currently, 500,000 people depend on the river for drinking water, including those residing in the communities of Lowell, Lawrence, Tewksbury, Methuen, and Andover, in Massachusetts and Nashua, N.H. Two more cities in New Hampshire—Manchester and Concord—are considering the river as another source of water.

In addition, the Merrimack is historically important as one of the birthplaces for industrialization, with the river powering textile mills and other factories. The region, "was then to industry what the Silicon Valley is today," said Theodore Steinburg, author of *Nature Incorporated: Industrialization and the Waters of New England*. Today, 847

The river has been a major asset on the North Shore for centuries, but those who work to protect it are concerned about residential development near the waterway. (*Wikipedia photo*)

dams harness the river and its tributaries, and some are candidates for removal to help restore river health and fisheries.

The Merrimack River is one of the three most important large rivers on the East Coast in its conservation value to migratory river herring and one of the six most important for 12 migratory fish species. The Merrimack watershed also supports at least 75 state and federally listed endangered species, numerous pairs of bald eagles, the largest tidal marsh habitat in New England, and a portion of the Atlantic Flyway bird migration route.

THE THREAT

Pavement is rapidly replacing trees across the Merrimack River watershed. The impact of unsustainable development on land, forests, habitat, and water quality is the largest threat that the Merrimack River watershed faces today.

The U.S. Forest Service ranks the Merrimack River watershed as the most threatened in the country due to the development of forest lands, fourth for associated threats to water quality, and seventh for loss of habitat for species at risk. An estimated 40 to 63 percent of the private forested land in the watershed is projected to be developed by 2030. These threats are the direct result of the growing population and associated development occurring throughout the watershed, especially in southern New Hampshire, the most rapidly developing part of the state. For example, the average population growth rate between 2000 and 2010 in the three main New Hampshire counties of the Merrimack River watershed (Belknap, Merrimack, and Hillsborough) is nearly two times greater than the New England average.

The textile industry has a long history on the Merrimack and on the Spicket River, shown here in about 1893. (*Lawrence Public Library*)

Protecting the extensive forests in the Merrimack River watershed is essential to preserve their natural filtration function, which removes nutrients, pathogens, and other pollutants in order to help clean the river's water. The looming loss of these forests threatens the water supply of the more than 500,000 people who depend directly on the Merrimack River for their drinking water today, along with the additional 200,000 residents expected to use the water in the near future.

Pollutants that flow through the Merrimack watershed ultimately impact the Gulf of Maine's marine ecosystem. In fact, the Merrimack is already the second greatest contributor of nitrogen and phosphorus to the Gulf of Maine. In the coming years, the National Oceanic and Atmospheric Administration predicts an increase in rainfall and flooding for New England rivers due to climate change, which will only increase polluted run-off. With the continued transformation of these vitally important forested lands into suburban developments, the ability of the land to absorb, hold, and naturally clean stormwater will be greatly reduced.

The loss of forested lands also reduces habitat for threatened and endangered species.

WHAT MUST BE DONE

In order to ensure clean drinking water and protect fish and wildlife in the Merrimack River from poorly planned development, the U.S. Environmental Protection Agency (EPA) needs to create a regional watershed team and implement key safeguards including protection for important forest lands along rivers and streams, green infrastructure solutions, and improved stormwater management to reduce the excess nutrients and pathogens in the river.

Scores of mills were built in Lowell, and many still stand today. (*Dyke Hendrickson photo*)

More than 500,000 people in the Merrimack Valley get their drinking water from the Merrimack, and so a clean river remains a high priority. (*Dan Graovac photo*)

The EPA has created watershed teams in other basins facing chronic water quality challenges including the Chesapeake Bay, Great Lakes, and the Gulf of Mexico. Such an approach empowers local citizens to take greater responsibility for river stewardship.

Currently, EPA is already working closely with the City of Lawrence, Merrimack River Watershed Council, the Commonwealth of Massachusetts, and other partners to improve Lawrence's water. Now is the time to take the action to the next level. The EPA has a once-in-a-lifetime opportunity to make a lasting positive impact on the Merrimack by facilitating and funding a plan, backed by local citizens, to conserve the river through smart planning and acclerated land protection while there is still time.

3

THE RIVER UNITES
A TEAM OF KAYAKERS

Derek Mitchell is executive director of the Lawrence Partnership and was among a group of community leaders who in 2019 kayaked down the Merrimack River to highlight environmental concerns. The trip drew attention to the river's importance as a recreational and economic asset in the region. The remainder of this chapter is an essay he wrote after the trip.

My life is intertwined with the Merrimack River. I drive across it at least twice a day. I look over its banks from the converted mills that line its shores. I live and work in the cities that were fueled by its waterpower. Yet, I rarely give much attention to the mighty Merrimack that runs from Franklin, N.H. all the way to the Atlantic Ocean.

This all changed when Lane Glenn (president of Northern Essex Community College) asked me if I wanted to paddle the full 117 miles of the river with him in a kayak. We set out with ten other collaborators to raise awareness about the recreational, economic, and environmental opportunities of this natural asset. But it is only now, nearly three weeks later, that I fully appreciate what I learned about those four days.

They are lessons that apply to the river but go far beyond that. We have much more in common than we realize. Our trip was full of differences—public, private and non-profit representation; Democrats and Republicans; men and women; residents of different towns and cities.

But when the paddles hit the water, we realized how aligned our thinking was about the river. More importantly, there was a recognition that our perceived differences ultimately made our coalition stronger. There are different state bills in the process that address the river, but at the end of the day, their authors (many of whom were in boats on the river) are all working in the same direction, despite representing different communities along the river.

There are more stools within reach than we sometimes realize. Far too often we believe problems are too big, too complex, and require too many resources to be dealt with locally.

Derek Mitchell, the author of this essay, says the kayak trip down the Merrimack was memorable.

While environmental stewardship of the river does require regional collaboration and the infusion of dollars, we overlook the depth of local expertise to be leveraged.

We were lucky to be joined by Gregg Coyle, the staff engineer of the Lowell Wastewater Treatment Facility, for the last two days of the trip, and we got a first-hand look at the monitoring equipment he has set up along the river.

More importantly, we learned about the cutting-edge technology that would allow engineers like Gregg to measure and monitor bacteria levels in near real-time for short dollars.

These small-scale investments, coupled with expert monitoring by folks like Gregg, could leapfrog us ahead in our understanding of water safety and contributed to policy and resource allocation to make the river even cleaner.

Our focus on deficits often—wrongly—overshadows our assets. We started the trip with a strong focus on combined sewer overflows (CSOs), the releasing of waste into the river during big rainstorms. As a result, we were all concerned when we paddled through Manchester—a major CSO contributor—the day after a big rainstorm.

But the conversation changed course the next night when we landed in Lawrence, to a community conversation on the dock of the Greater Lawrence Boating Program, and a question was raised about cleanliness. A fellow voyager jumped on the opportunity to remind us all that the river (that we had traveled on) had been incredibly clean, and the wildlife beautiful and the rope swings a lot of fun. It was an assertion that could not have been made a generation ago.

The desire of so many of us to get on the river for four days was an indication of how appealing the river is in its current form.

Boats of all sizes and shapes frequent the harbor in Newburyport.

Shared experience is a unique way to align, motivate and activate. Prior to the trip I knew most of the folks involved and liked all the people I knew. On the back end of the four-day voyage, which included as many blisters and inner prayers for some current in our favor as did laughs and moments of inspiration, I now consider all of them comrades.

While that was not the intent of the trip, I would argue it was the most gratifying byproduct. Our bond is born out of not only shared adversity and experience but also the mutual commitment to engage in such a trip in the first place, and willingness to do so on behalf of the greater good.

While the trip was decidedly about the river, in support of the river and on the river, the Merrimack was really just the stage for a broader set of truisms. The trip was a reminder of the linkages that connect the disparate communities that have both historic and geographic ties.

Ours is a shared regional fate, and our investments in the Merrimack River—and recognition of it as an asset worthy of investment—are proxies for our investments in a shared future. Keeping in mind the local tools and knowledge at our disposal, our aligned interests and our foundation of assets to work from, we are well-positioned to be effective stewards of the river—and all it represents.

As the river has become cleaner following the Clean Water Act of 1972, more recreational vessels have come to Newburyport.

The Merrimack in Amesbury is a fine place to keep a boat. (*Dyke Hendrickson photo*)

4

NEWBURYPORT:

THE JUNCTION OF THE MERRIMACK AND THE ATLANTIC

Newburyport is at the end of the line when it comes to the river churning 117 miles southeast from the hills of New Hampshire into the Atlantic Ocean. Yet it is first when it comes to "the start" of the waterway. Its history reflects that it developed from a major shipbuilding and trading town to its current status as a fashionable suburb and tourist city.

Today, it serves as a destination for thousands, though there are no major motels in the community. Some visitors stop here on their way to New Hampshire, Maine, or the Maritimes. It offers a long boardwalk along the river, where visitors can amble, power-walk, or just loiter. Close to 1,500 recreational boats are registered here.

In addition, it has several miles of "rail trail" where folks can leave the river to wander through the city's handsome neighborhoods. Also, Cashman Park along the river offers a playground, a basketball complex, a tennis court, a soccer field, and an active boat-launching operation.

Samuel de Champlain discovered the river in 1605. It was not called Merrimack at that time, but explorers soon determined that it ran deep into the area later known as New Hampshire. For many years, it was spelled either "Merrimac" or "Merrimack." The spelling was formalized in the state legislature in 1914 to Merrimack, with a "K."

Newbury was founded in 1635 by English colonists. Almost from its inception, this community was recognized as one that was made up of disparate denizens: farmers and "water-siders." The two vocational groups had different interests and aspirations. In 1764, residents voted to let the water-siders start their own town: Newburyport.

It must have been a dramatic time for debate in this community. Many colonists were outspoken about resisting England's strong controlling hand. Yet those in the Newbury area were arguing about a different kind of separation. In addition to discussing revolution, many residents near the river wanted to divorce themselves from Newbury as well. One guesses that the town meetings of the day were loud and clamorous. Separation occurred in 1764, and this "new" Newburyport emerged as a major community in New England.

Newburyport was a dynamic shipbuilding community in the age of sail and international trade, as this painting by Richard Burke Jones suggests. It depicts a scene from the 1850s.

Newburyport developed into a leading shipbuilding town. The historian Samuel Eliot Morison wrote in his *Maritime History of Massachusetts*:

The lower Merrimack from Haverhill to Newburyport was undoubtably the greatest shipbuilding center of New England. Close to 1115 vessels were constructed and registered there between 1793 and 1815.[1]

Newburyport historian John J. Currier wrote, "All the cordage, sails, blocks, pumps, ironwork, anchors and other fittings were made locally, employing hundreds of skilled mechanics." Full employment followed. The local economy thrived.

Many of the great mansions of High Street were built around 1790 to 1811. The Embargo Act of 1807 slowed development, and the Great Fire of 1811 decimated half of the town and its commerce. Yet for two decades and more, this part of the Merrimack Valley was thriving.

Newburyport prospered on shipbuilding, in part because of its access to timber. Forests lay on both sides of the Merrimack River for up to 100 miles, and there was rarely a lack of timber. Dams and canals were built in Lowell and Lawrence in the nineteenth century to allow for commercial movement. Still, timber continued to float down the river.

The docks at Newburyport represented a center of international trade. Ships and crews from Newburyport were as courageous as the astronauts of today. Here is a description from the local historian, John J. Currier, in 1906:

From 1784 to 1794, there were many vessels arriving in Newburyport from Guadeloupe, Port au Prince, St. Martins and Suriname with cargoes of molasses, sugar, coffee and cotton. Also, there were ships from Madeira with wine, from Turk's Island or Cadiz with salt, from Ireland with linen, from Rotterdam with gunpowder, from Dunkirk with earthenware and carpeting, from Bilbao with silk handkerchiefs, silk gloves and glassware.[2]

Merchants in Newburyport exported lumber, fish, and ice, and customers in the West Indies were active buyers. They imported sugar and molasses, and the town was one of the leading makes of rum in the region.

In addition to developing a local fleet, shipbuilders sold vessels to private shippers of other nations. Many a vessel built in Newburyport was utilized to carry slaves. Importation of slavery to the U.S. ended in 1808, but vessels were built long after that for slavers going to the West Indies and South America.

Merchants in Newburyport were patriotic, and they sometimes paid for it. Currier said that from 1775 to 1783, the local shipowner Nathaniel Tracy "was the principal owner of 110 merchant vessels. Of this total only 13 were left at the end of the war, all the rest taken by the enemy or lost." Tracy, though, took many enemy vessels himself and he always had a fleet back at the dock. He lived the final years of his life in what is now the Newburyport Public Library on State Street.

Perhaps because Newburyport was a vital center during the Revolutionary War, national leaders steered business its way following the American Revolution. In 1789, President George Washington and Secretary of the Treasury Alexander Hamilton determined they needed to generate money for the national treasury by collecting tariffs in ports along the Atlantic. They also wanted to stop illegal smugglers who avoided paying any tax.

Hamilton was a native of the West Indies, and he knew much about international trade; he was also aware of the methods that captains used to avoid paying. Ten "revenue cutters" were commissioned to be built for the purpose of patrolling harbors and collecting money. Two of the ten were built in Newburyport, and the first cutter, the *Massachusetts*, was constructed here. One hesitates to suggest that there was a political scent to this contract, but Washington himself visited the community at this time. He was acquainted with many Newburyport leaders.

The captain of the *Massachusetts* was John Foster Williams. Today, the Boston headquarters of the Coast Guard's District 1 headquarters (New England, New York, northern New Jersey) is named after this mariner.

The revenue cutters eventually became the basis of the U.S. Coast Guard, and Newburyport is considered the birthplace of the Coast Guard because of its production of the *Massachusetts*. The term "Coast Guard" was first applied in 1915. At that time, the Revenue Cutter Service merged with the Life-Saving Service. In 1939, the Lighthouse Service became part of the organization, and the three units together emerged as the modern Coast Guard. In 1965, President Lyndon Johnson signed a decree naming Newburyport as the birthplace of the Coast Guard. That declaration can be found at the Custom House Maritime Museum adjacent to the Merrimack.

Newburyport was also a leader in building ships for the nascent U.S. Navy in the 1790s. The *Merrimac* was built in this community in 1798. Its builders were adept and energetic. Skilled tradesmen made sails, rigging, rope, planking, and metal products. Carpenters did excellent work on ship interiors. Their presence in the community was one reason why so many residences are handsomely designed and well-built.

The Symington was one of the last large ships of sail launched in Newburyport (1893). (*Newburyport Archival Center of the Newburyport Public Library*)

Before they had cars, residents loved a trip on a boat. (*Newburyport Archival Center of the Newburyport Public Library*)

Their work drew attention throughout the country. In 1853, the swift clipper ship *Dreadnought* was built here and went on to set a record for the fastest crossing of the Atlantic—nine days. There is a stone monument today at the base of Ashland and Merrimac streets in Newburyport that notes this once-famous achievement.

Besides shipbuilding and merchant trade, fishing was a key industry. Men entering the fishing trade were the uneducated and unskilled, in part because this was a very dangerous occupation that did not pay well. The leading young men in the community went into shipbuilding, trans-Atlantic trade, or "counting houses" that supervised commerce.

So, the lesser lights went onto small fishing vessels. Some worked locally but many went far into the Gulf of Maine. Boats were frequently lost, and it was said that there were more orphans here than in most inland communities. During the "Yankee gale" in 1851, ninety-two fishing vessels from the North Shore were destroyed in a sudden Atlantic storm. Twenty-four of the craft were from the Merrimack River in Newburyport, and their loss meant misery and destitution for numerous families.

Perhaps because of the large numbers of orphans, Newburyport had the first tax-supported "female high school" in the country. Other communities offered courses and regimes in dance, language, and domestic arts, but this community had the first tax-supported public high school for girls, which ran from 1844 to 1866. It was terminated because of consolidation and was lost to a fire several years after closing.

Newburyport became a leading port despite having a very narrow and shallow harbor. That circumstance still exists today, and many recreational boaters who are new to Newburyport find themselves halted on sandbars.

Newburyport offers "islands in the stream" just upriver from the city. (*Bryan Eaton photo*)

The entrance has been altered over the years. In 2015, crews fortified both the north and south jetties. Each federally funded jetty cost about $10 million to construct. Improved navigation was the purpose. The jetties were constructed to keep shifting sand out of the harbor, but photos indicate that a sand bar forms each year. Recreational boaters not familiar with the harbor often run aground at low tide.

Records at the Custom House Maritime Museum reflect that scores of wrecks occurred off Plum Island and Salisbury Point over past centuries. Even on good days, captains would have to fight a strong river current while having to deal with the tide, and they were sailing through a section of the river that might not be deep enough for the ship, depending on its weight and the tide. Today, the depth of the waterway can get as shallow as 16 feet, according to Newburyport harbormaster officials. One of the reasons the city's shipbuilding trade declined in the late nineteenth century was that ships got too large for the harbor. A ship of sail weighing 500 tons or more might be able to leave at high tide, but it could not return weighed down with cargo.

Another reason Newburyport lost its prominence was the arrival of the Middlesex Canal in 1804. The highly engineered waterway ran from modern Lowell to Charlestown. Many merchants chose to send their goods to the Boston area by this route. When the textile revolution began in Lowell in the 1820s, much of that trade bypassed Newburyport.

In Newburyport, one of the last large wooden ships of sail, the *Symington*, was built here in 1893. Newburyport shipbuilders did not transition into the construction of metal steam vessels in the late nineteenth century, in part because they were too heavy to traverse the shallow harbor.

The river in Newburyport also served as a major source of rest, relaxation, and entertainment. Few families could buy their own boats in the nineteenth century, and this was well before the era when families had an automobile. So men, women, and children who could afford a day on the water often traveled to the Merrimack, sometimes to bathe and, more often, for scenic rides aboard commercial craft.

In the nineteenth century, before most families had showers and bathtubs, some residents sought respite from the summer heat and grime in riverside bathhouses. As early as 1807, a bathing house was created off Merrimac Street and offered "hot and cold baths, any day of the week, from 6 a.m. to 10 p.m." In 1827, a bathing house was created on "Jackson wharf," and in 1834, there was a bathing house on Merrimac Street, near the bottom of Strong Street, open from 5 a.m. to 10 p.m. "for the use of ladies and gentlemen desiring hot or cold baths."

In 1845, records show that another bathing house, "nearly opposite the James Steam mills," opened "with all the modern conveniences and well patronized during the summer months," according to historian Currier, in *History of Newburyport, 1764–1909, Volume II*, printed in 1909. In 1892, the City Improvement Society presented the city with a floating bathhouse, which was moored near the dock at the foot of Winter Street, and open free of charge in August of that year.

In terms of a day on the water, small steam-driven boats arrived in number on the North Shore in the mid-nineteenth century, and numerous companies offered services on the river in that era. By the end of that century, small steam yachts and excursion boats were available for riders.

The next year, the *Three Brothers* vessel was advertised to make frequent trips from Newburyport to Plum Island and Salisbury Beach. In 1880, two small steamers, the *White*

Fawn and the *Wanderer*, "made trips almost daily from Newburyport to Gloucester, Isle of Shoals and Portsmouth."

The *Minneola*, a twin-screw steamer, built in Newburyport in 1887, made frequent trips from Haverhill to Boar's Head, Portsmouth, and the Isle of Shoals.

The *City of Haverhill*, a side-wheel steamer that belonged to the Haverhill, Newburyport, and Boston Steamboat Company in 1902, also had a regular route to Boston.

Perhaps the most famous was the *Dora*, built by Frederick S. Moseley in 1895 for private use. The Moseleys were among the wealthiest families in the region, and they often moored the craft on the Merrimack across from their home, now the site of Maudslay State Park.

Currier says it made frequent trips during the next three or four years to places of interest in the vicinity of Newburyport: "She was a swift and staunch sea-boat, able to withstand the fury of the winds and waves, and provide ample accommodations for comfort and convenience."

The *Dora* was sold to the U.S. Government in 1899 and was subsequently employed in the hospital service near Cuba, perhaps dealing with infirmities resulting from the Spanish–American War. One of the last pleasure boats that might be recognized as a transitional craft between "the old days" and modern pleasure vessels was the steamboat *Sabino*, a wooden-passenger boat built in Maine in 1908.

After many years serving the islands and waterways of southern Maine, it was restored by the Corbin family of Rings Island and plied the Merrimack in the 1960s and early 1970s. The *Sabino* was the last remaining wooden, coal-fired steamboat in operation in the U.S., The *Sabino* was purchased in 1974 to serve as a working exhibit at the Mystic Seaport Museum in Connecticut. It was formally designated a National Historic Landmark in 1992, perhaps reflecting the nostalgic tone of those who looked back on that era of transportation.

In the late nineteenth century, recreational boating became more popular for local families here. *About Yachting 1894: A Yachting Souvenir of the American Yacht Club* notes that the American Yacht Club was organized in March 1885 at a "party of gentlemen" assembled at the Brown Square House. Twenty-five members were enrolled at this meeting.

In the summer months, members occupied the clubhouse on Coffin's Wharf, and in the winter, their rooms were on State Street. Membership grew, and members would sit on the piazza of the club, watching sailing boats glide back and forth on the harbor.

A second club further up the river soon appeared. Local historian Jean Foley Doyle notes that the North End Boat Club was established in 1895, as a club a corporation for the purpose of encouraging athletic sports and yachting.

The AYC was the retreat of those who wanted to sail. Doyle said, "The North End Boat Club, however, was destined to become a motorboat association formed almost entirely by working-class men from their section of the city."[3] Both clubs are still active today, though each has been rebuilt or remodeled after fires, adverse weather, or other reasons.

Today, numerous craft still cater to those who want to be underway on the river. There is a vessel for whale-watching, one for educational tours, several for youngsters learning about the seaside environment, a catamaran for sailing trips, and several craft that take fishermen to the sea. Newburyport city officials say close to 1,500 craft are registered each year at the harbormaster's office. Close to 500 more are registered in Salisbury, Amesbury, Groveland, Merrimac, and Haverhill. All are carrying on the venerable tradition of taking landlubbers out on the historic Merrimack.

Coast Guard vessels on the river are a common sight in Newburyport, which is the birthplace of the U.S. Coast Guard (1790). (*U.S. Coast Guard photo*)

PARKS ALONG THE MERRIMACK

Newburyport is blessed with many parks along the Merrimack River. The following are some that can be visited for free.

Cashman Park

This park off Merrimac Street is the busiest public park in the state for launching boasts. The cost to do so is minimal, and there is plenty of parking. The original name of this riverfront park was Central Park. The city administration of Mayor Michael Cashman took land by eminent domain starting in the early 1920s. Prior to Cashman's initiative, the parcel lay between a landfill, a distillery, and Towle Manufacturing Company. One of the park's most popular uses in the 1920s and '30s was baseball. Local teams would play before as many as 3,500 fans. It was dedicated to Cashman in 1944.

Hale Memorial Park

This 2-acre park is located on Water Street between Madison and Marlborough streets. There is not much public parking, but it is a quiet sedentary retreat with benches and manicured lawns. It is named after Joseph. W. L. Hale, who died in 1960 and left acreage for a Hale Memorial to honor his ancestors, who arrived in 1637. He graduated from Newburyport High School and MIT; he was a professor at Pennsylvania State University. He was an inventor and also wrote several textbooks relating to mechanics and engineering.

Joppa Park is one of many public riverside parks in Newburyport. (*Dyke Hendrickson photo*)

Joppa Park

There used to be much clamming in the Joppa area. Today, the bacteria levels sometimes fall enough to permit gatherers to dig their quota. A clam-purification plant operates on the north end of Plum Island, close to the river itself. Joppa is located on the Merrimack River at the south end. In the eighteenth and nineteenth centuries, it held small homes and sheds that hosted clamming and fishing operations. In 1968, Mayor Byron Matthews initiated the building of a sea wall to avoid the constant cleanups of river debris flooding onto Water Street after storms. Construction of the sea wall was completed in 1972. This area, bordered by the sea wall on one side and Water Street on the other, was redeveloped into Joppa Park in 1971. It is a popular site from which to launch a kayak.

Moseley Woods

This 18-acre park, once known as "The Pines," is at the corner of Merrimac and Spofford streets and is on the banks of the Merrimack. It is a valuable asset for locals who needed a place to walk their dogs. It was left to the city in 1922 in the will of Charles W. Moseley. It is owned by the city and the Moseley Trust; it offers trails, picnic tables, grills, and a playground as well as scenic views of the Merrimack.

Maudslay State Park

This park is a gem. The park covers approximately 450 acres on the river. Though boating and swimming are banned because of high bacteria counts from upriver, it is a fine park for hiking and horseback riding. A popular "art in the park" event is hosted in summer. In spring, it offers acres of gorgeous bushes and flowers. In winter, cross-country skiers flock to the area. The main channel of the estuary runs beneath the bluffs of the park. The channel is navigable to small craft and is marked by buoys.

The state park was created from the early-twentieth-century estate of Frederick Strong Moseley, the son of Edward Strong Moseley (1813–1900), a prominent citizen of Newburyport. Moseley is a variant of Maudesley or Maudesleigh, an English name going back centuries in that country.

From the park, one can often sea eagles. The return of the eagles is related to the return of their food supply. By 1950, the Merrimack River was for the most part devoid of healthy marine life due to chemical effluents from the cloth and paper mills upstream as well as the dumping of raw sewage into the river from communities on its banks. Since the Clean Water Act, the quality of the river has improved. With climate change, much of the river is free of ice in winter. This enables eagles to grab fish for food.

In 1985, the state of Massachusetts began administration of the acreage. A parking lot for close to 100 vehicles makes it a destination site. Horse trailers are permitted, and riders are frequently seen in the park.

Notable wildlife includes harbor seals, which are often found on Badgers Rock in the Merrimack River in the fall and winter. Birds that can be sighted include black ducks, green-winged teal, and great blue herons, along with pectoral, solitary, and least sandpipers.

In recent years the Merrimack River Beach Alliance has monitored the health of the river and nearby Plum Island. Sen. Bruce Tarr, R-Gloucester, and Salisbury leader Jerry Klima have led the efforts to raise close to $20 million in federal funds to create new jetties. Newburyport Mayor Donna Holaday has been a leader in pushing for a warning system that will inform boaters, fishermen, and riverside residents when there has been a CSO upstream.

5

SALISBURY:
A SEASIDE TOWN KNOWN FOR SUMMER FUN

Many adults who grew on the North Shore of Massachusetts have fond memories of Salisbury Beach. Some walked along the Merrimack or swam in the Atlantic. Others parked their cars and camped at the state park there. Thousands thrilled themselves on the rides or gorged themselves on pizza and cotton candy. The old hotels and semi-dangerous rides are gone, but Salisbury remains one of the oceanside attractions of New England.

Salisbury was once the territory of the Pentucket tribe of Pennacook Indians. As occurred in countless other communities in North America, many Native Americans died of disease brought by white settlers.

As Jared Diamond wrote in *Guns, Germs and Steel*, Europeans at this time were immune from many communicable diseases. Their ancestors had suffered through them centuries earlier, but the Indians were not immune. Many more Native Americans died of disease than they did in the limited but well-publicized warfare that took place in New England.

Salisbury was settled by the English in 1638 as Colchester and incorporated in 1640 as Salisbury, after Salisbury in Wiltshire, England. In 1866, Beach Road was constructed across Great Marsh, providing access to the town's 5 miles of smooth, hospitable beach. It developed into a thriving summer resort, lined with hotels, restaurants, shops, cottages, arcades, and amusement parks.

It is one of the few communities on the Atlantic whose modern history is built around the development of amusement attractions. A carousel called the Flying Houses was installed in 1914. Later, the Sky Rocker, the beach's first rollercoaster, was created. A dodgem ride was installed in 1920 and lasted until about 1980. The boardwalk always included a venue for entertainment, and major entertainers in the twentieth century included Glenn Miller, Ella Fitzgerald, Louis Armstrong, Frank Sinatra, and Liberace.

The resort remained vibrant through the 1960s, but forms of gentrification began in the 1970s. The Wildcat, the last roller-coaster, was removed in 1976. Pirate's Fun Park, the last small amusement park, closed in 2004 to be replaced with condominiums.

A famous landmark in Salisbury is Ben Butler's Toothpick. Butler was a governor of Massachusetts.

Yet a great beach is a great beach. Thousands come every summer to take rental units near the Atlantic. The 355-acre Salisbury State Park offers 484 camping berths and several boat launches. Like many resort communities on the North Shore, city leaders are constantly worrying about having enough sand to draw tourists, and that brings the reader to the subject of jetties.

In recent years, the Army Corps of Engineers has taken the lead in designing and supervising the revamping of the jetties. The Corps of Engineers' key motivation is navigation. They want the Merrimack to be as deep as possible so that pleasure craft can enter and depart. The thinking is that the jetties will limit the movement of sand. They hope the development of sand bars will be kept to a minimum.

The north and south jetties were shored up between 2014 and 2018 at a federal cost of about $20 million. Navigation improved somewhat, but many business professionals in Salisbury say that their presence has denuded part of the beach of some sand. This is not a trivial point. If owners of rental properties cannot provide an enticing sand beach, they will lose customers, and if visitors to the music hall cannot get a majestic view of sand and surf as the sun sets and the wine is poured, their customers could drift off to other communities.

The jetties harnessing the Merrimack River also affect Newburyport and Newbury. The jetties on the south side of the river have influenced the aggregation of sand in those communities. Some areas of beach near the "center" of Newbury have lost a great deal of sand.

One result is erosion, and several houses along the beach in Newbury have fallen into the Atlantic over the last decade. Structures in Newburyport, farther from the high-tide mark, are sometimes flooded.

The arcades in Salisbury have been enjoyed for years.

Salisbury Beach, at the juncture of the Merrimack and the Atlantic, is one of the most beautiful beaches on the North Shore.

Half of Plum Island is in Newbury, and half is in Newburyport. The two communities sometimes have different views on how to utilize the island. Newbury, which needs real-estate tax dollars, is amenable to approving building permits. Backhoes are often seen digging into the dunes to make way for another structure. In contrast, Newburyport stresses enhancing and preserving the dunes. Schoolchildren sometimes spend the day there putting in dune grass and small bushes, so the sand will not drift away.

It appears that the Merrimack River gives, and it takes away. The creation of jetties is part of an ongoing effort to preserve beaches in Salisbury, Newburyport, and Newbury. Also, when communities outlawed wells shortly after 2000, sewerage pipes were sent along the river to collect the waste from the island. This meant that larger houses could be built. Now, most structures built along the river or on Plum Island offer at least 3,000 sq. feet and cost more than $1 million.

On the subject of river-borne curiosities, one must recall an enormous gaffe brought by managers of the sewage-treatment plant in Hooksett, N.H. In 2011, the managers tried an experimental plan for dealing with sewage solutions.

This method involved utilizing plastic disks the size of a silver dollar to control effluent, but an oversight occurred, and the plant was flooded. The entire collection of water-cum-sewage was released from the plant. The disks—about 4.3 million of them—were released in the river. For months, they were found along the Merrimack. News reports have indicated some have been picked up on the shores of France and Great Britain. They are still found along the Merrimack in 2020.

Salisbury remains a very popular destination for families looking for an inexpensive vacation. Campsites along the river are full almost every summer. One well-known "landmark" in the river off Salisbury is Ben Butler's Toothpick. Butler (1818–93) was a general in the Civil War and later a congressman and governor of Massachusetts. He lived in Newburyport for a time. He loved sailing. When this navigational aid was constructed in about 1873, it was named after Butler. On the topic of ports, Butler was despised in New Orleans, which he supervised during the war after it fell to the Yankees.

6

AMESBURY:
A COMMUNITY WITH A VIBRANT HISTORY

Parts of Amesbury have some of the most scenic views of the Merrimack. The drive along Main Street and past Hatter's Point offers gorgeous vistas not only of the river but of the virgin forest of Maudslay State Park across the water. The presence of historic Lowell's Boat Shop hugging the shore—and still making wooden vessels—suggests that this was once one of the most prominent communities of the lower Merrimack Valley.

Beginning as a modest farming community, it developed an aggressive maritime and industrial economy. Shipbuilding, shipping, and fishing were important. The ferry across the Merrimack to Newburyport was a lively business until the construction of bridges. Narratives at the riverside Alliance Park note that Amesbury was a major shipbuilder in the late eighteenth century.

The Powwow River was always a major force in the community. In the nineteenth century, textile mills were built at the falls, as was a mechanized nail-making factory, believed to be among the nation's first. The Merrimack Hat Company was a national leader in that field. Beginning in 1853, Amesbury became famous for building carriages, a trade that evolved into the manufacture of automobile bodies. Prominent manufacturers included Walker Body Co., Briggs Carriage Co., and Biddle and Smart. The industry ended with the Depression.

In 1876, the town of Merrimac separated from Amesbury. In 1997, Amesbury changed its status to a city and adopted the mayor and municipal council form of government. Much municipal activity in recent years has been focused on rebuilding and revitalizing the downtown area. The old mill buildings are charming and much of the residential architecture is admirable. The Powwow runs through it. Above all, its scenic location on the Merrimack River has charmed visitors for decades.

A massive $292 million federally funded bridge linking Amesbury to Newburyport on I-95 was recently dedicated in the name of poet John Greenleaf Whittier. Whittier appreciated the value of the river and his home grounds. He was born near Haverhill, and his work has always focused on the Salisbury, Amesbury, and Haverhill areas.

The Whittier Bridge links Newburyport and Amesbury on Route 95. (*Dyke Hendrickson photo*)

Bucolic scenes such as this are common on the Merrimack. (*Bryan Eaton photo*)

The following is one of his most famous works, and the early spelling is used for the river.

THE MERRIMAC

John Greenleaf Whittier, 1841

STREAM of my fathers! sweetly still
The sunset rays thy valley fill;
Poured slantwise down the long defile,
Wave, wood, and spire beneath them smile.
I see the winding Powow fold
The green hill in its belt of gold,
And following down its wavy line,
Its sparkling waters blend with thine.
There's not a tree upon thy side,
Nor rock, which thy returning tide
As yet hath left abrupt and stark
Above thy evening water-mark;
No calm cove with its rocky hem,
No isle whose emerald swells began
Thy broad, smooth current; not a sail
Bowed to the freshening ocean gale;
No small boat with its busy oars,
Nor gray wall sloping to thy shores;
Nor farmhouse with its maple shade,
Or rigid poplar colonnade,
But lies distinct and full in sight,
Beneath this gush of sunset light.
Centuries ago, that harbor-bar,
Stretching its length of foam afar,
And Salisbury's beach of shining sand,
And yonder island's wave-smoothed strand,
Saw the adventurer's tiny sail,
Flit, stooping from the eastern gale;

And o'er these woods and waters broke
The cheer from Britain's hearts of oak,
As brightly on the voyager's eye,
Weary of forest, sea, and sky,
Breaking the dull continuous wood,
The Merrimac rolled down his flood;
Mingling that clear pellucid brook,
Which channels vast Agiochook,
When spring-time's sun and shower unlock
The frozen fountains of the rock,

And more abundant waters given
From that pure lake, "The Smile of Heaven,"
Tributes from vale and mountain-side,
With ocean's dark, eternal tide!

On yonder rocky cape, which braves
The stormy challenge of the waves,
Midst tangled vine and dwarfish wood,
The hardy Anglo-Saxon stood,
Planting upon the topmost crag
The staff of England's battle-flag;
And, while from out its heavy fold
Saint George's crimson cross unrolled,
Midst roll of drum and trumpet blare,
And weapons brandishing in air,
He gave to that lone promontory
The sweetest name in all his story;

Of her, the flower of Islam's daughters,
Whose harems look on Stamboul's waters,
Who, when the chance of war had bound
The Moslem chain his limbs around,
Wreathed o'er with silk that iron chain,
Soothed with her smiles his hours of pain,
And fondly to her youthful slave
A dearer gift than freedom gave.
But look!—the yellow light no more
Streams down on wave and verdant shore;
And clearly on the calm air swells
The twilight voice of distant bells.
From Ocean's bosom, white and thin,
The mists come slowly rolling in;
Hills, woods, the river's rocky rim,
Amidst the sea-like vapor swim,
While yonder lonely coast-light, set
Within its wave-washed minaret,
Half quenched, a beamless star and pale,
Shines dimly through its cloudy veil!

Home of my fathers! I have stood
Where Hudson rolled his lordly flood:
Seen sunrise rest and sunset fade
Along his frowning Palisade;
Looked down the Appalachian peak
On Juniata's silver streak;
Have seen along his valley gleam

The Mohawk's softly winding stream;
The level light of sunset shine
Through broad Potomac's hem of pine;
And autumn's rainbow-tinted banner
Hang lightly o'er the Susquehanna;
Yet wheresoe'er his step might be,
Thy wandering child looked back to thee!
Heard in his dreams thy river's sound
Of murmuring on its pebbly bound,
The unforgotten swell and roar
Of waves on thy familiar shore;
And saw, amidst the curtained gloom
And quiet of his lonely room,
Thy sunset scenes before him pass;
As, in Agrippa's magic glass,
The loved and lost arose to view,
Remembered groves in greenness grew,
Bathed still in childhood's morning dew,
Along whose bowers of beauty swept
Whatever Memory's mourners wept,
Sweet faces, which the charnel kept,
Young, gentle eyes, which long had slept;
And while the gazer leaned to trace,
More near, some dear familiar face,
He wept to find the vision flown,—
A phantom and a dream alone![1]

John Greenleaf Whittier was a major figure in American literature in the nineteenth century, and his poems included an ode titled "Merrimac."

7

The River was Key to Development in Merrimack Valley

As this narrative continues its journey upriver, the reader nears communities whose industrial identities were developed with the aid of the prodigious power of the Merrimack River. Massachusetts mill communities significantly influenced by the river include Haverhill, Lawrence, and Lowell. The National Park Service is an agency with authority on the river, and its historic park in Lowell is nationally known for adroitly recreating the history of the waterway.

This brief history was produced by the National Park Service:

The Early Merrimack: The Pennacook Confederacy

Pennacook Indians were this area's first recorded inhabitants. Each spring, tribes met at the Pawtucket Falls [in Lowell] on the banks of the Merrimack to fish during the day and conduct business at night.

Marriages and trade agreements between tribes strengthened the confederacy. At the close of the fish runs, tribes moved upriver to plant and harvest crops. In the seventeenth century, smallpox, war, and English settlement devastated the Pennacooks. Many died, while others moved north to Canada. By 1725, land once home to the Pennacooks was absorbed by the town of Chelmsford [now Lowell]. A way of life was gone forever.

The colonists and their descendants used the river for fishing and transportation. Merchants from Newburyport built the Pawtucket Canal (1796) to bring timber products around the Pawtucket Falls. Boston traders operating in the Merrimack Valley built the Middlesex Canal (1803) to ship goods to and from Boston. In 1814, a group of Boston merchants financed the country's first "integrated," water-powered textile mill. Located on the Charles River in Waltham, this mill housed all the machinery needed to turn raw cotton into finished cloth.

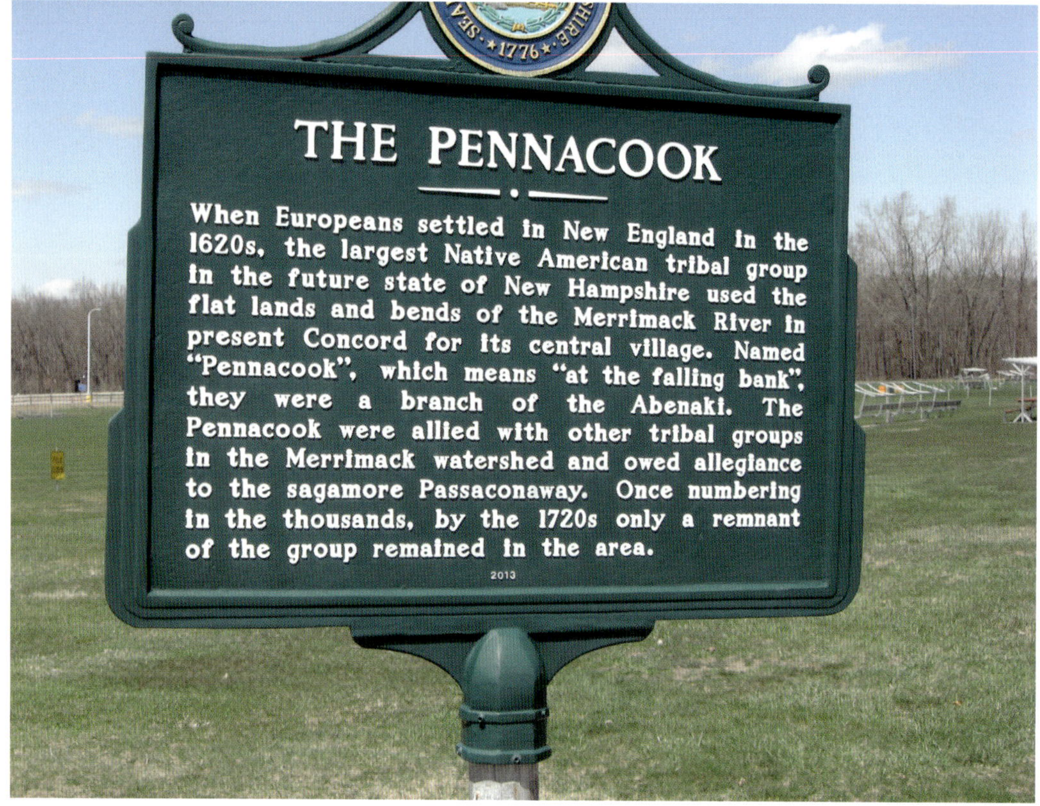

This sign in Concord, N.H., relates a grim story about Native Americans in New England.

In 1823, wishing to expand production, the Boston group began textile operations in East Chelmsford (now Lowell), using the 32-foot Pawtucket Falls. They founded 10 textile companies in the town they renamed Lowell. For the next 30 years, the city led the nation in cotton textile production. The giant textile companies viewed water only as an element of production. The amount of water reaching a mill was artificially controlled to maximize output.

Engineers controlled the water in the canals by building small, covered dams with gates that could be raised or lowered to release more or less water. Large dams on the river itself were another part of the system. Dams helped mill owners obtain a steady water supply by holding back water overnight. Lowell's dam, built just above the falls, turned the river into an 18-mile-long millpond. Other dams were built in Lawrence, Manchester, N.H., and elsewhere along the Merrimack.

Control of the Merrimack's sources in New Hampshire was also crucial to the water delivery system. The Lowell and Lawrence mills joined forces in the 1840s to buy rights to the waters of Lake Winnipesaukee, Newfound Lake, and Squam Lake. The growth of Lowell and other industrial cities on the Merrimack dramatically changed the river's ecosystem. One of the most noted changes was the disappearance of fish.

One of the first travelers to be concerned about the loss of wildlife was Henry David Thoreau. He wrote *A Week on the Concord and Merrimack River* in 1839, and it was published in 1849. In an expansion of his sentiments noted earlier in this book, he expressed concern even then:

> Salmon, shad and alewives were formerly abundant here … until the dam … and the factories at Lowell, put an end to their migrations hitherward…. Perchance, after a few thousands of years, if the fishes will be patient, and pass their summers elsewhere … nature will have levelled [*sic.*] … the Lowell factories, and the Grass-ground River [will] run clear again.[1]

In addition to building dams, textile makers dumped wastes into the river. The canals of Lowell were especially loaded with trash and dyes, making the water unfit for drinking. Population growth also contributed to the pollution of the water supply. By the 1870s, human waste and run-off had contaminated Lowell's wells, so the city began using filtered river water. Meanwhile, city sewage systems dumped wastes directly into the Merrimack.

In the 1920s, 12 million gallons of Lowell sewage entered the river each day. Sometimes diseases traveled downstream from one river city to another. Since the 1970s, the river's health has been improving, in part due to activism and legislation. The Clean Water Act of 1972 mandated waste treatment plants and regulation of toxic discharges. Progress has sometimes been slow: Manchester, New Hampshire, stopped regularly dumping raw sewage only in 1992.

The Anadromous Fish Conservation Act of 1965 led to a joint state-federal effort to restore migratory Merrimack fish such as salmon and shad. Salmon are taken from Lawrence to Nashua to spawn in a hatchery and are later released back into the river. Both wild and domestic salmon are stocked at various life-cycle stages. The goal is to get 3,000 salmon beyond Manchester [The historic salmon population was 30,000].

Shad are brought from Lawrence and released above Lowell to continue their journey. Fish ladders and elevators facilitate the migration of these fish. Other Merrimack fish include alewives, herring, and eels. The Merrimack still supplies power via six hydroelectric dams on the river and almost 100 small power projects. Hydropower is generated in Lowell by a plant on the Northern Canal and by turbines in former mills. The Merrimack is the second-largest surface drinking water source in New England. It serves hundreds of thousands of people through four water treatment plants.

The river is still beset by pollution problems. Salt, grease, trash, and pesticides run off into the river from cities and suburbs alike. Activist groups allege that poor monitoring allows violation of toxic regulations. Recovering fish species must still be stocked. Aging treatment plants need updating. Riverside vegetation buffers are often lacking. And, as in many other waterways, mercury is found in the Merrimack. The most stubborn problem now is Combined Sewage Overflows (CSOs). Normally, sewage treatment plants treat both run-off and sewage. However, heavy rains can cause such an excess of water that some waste goes into the river untreated, periodically lowering water quality.

Despite such problems, the river is fine for boating. Also, Lowell reopened its river beach in 1996, providing the state's only official river swimming east of Worcester. River fish are often suitable for eating, within state guidelines.

Wooden boats are still crafted at Lowell's Boat Shop in Amesbury. (*Bryan Eaton photo*)

Many parts of the Merrimack offer scenic views and recreational opportunities. (*Dyke Hendrickson photo*)

8

HAVERHILL IS USING RIVER AS CENTERPIECE FOR IMPROVEMENT

Many riverside communities are in a stage of rebirth in the twenty-first century, and Haverhill is making conspicuous gains.

A new edifice housing a branch of the University of Massachusetts Lowell (UMass Lowell) rises adjacent to the waterway. Apartments are being developed. Several banks and businesses have also put up buildings, and numerous new restaurants have added to the dining and nightlife. A riverside park is perfect for walking. The city is making a comeback and using the river as its centerpiece.

Haverhill was founded in 1640 as a frontier settlement by twelve Puritans from Ipswich and Newbury. Pentucket, which means "beside the winding river," was the original name of the settlement. Two years later, the "Haverhill Deed of Township" was signed, and Pentucket was renamed Haverhill after Haverhill, England.

On March 15, 1697, Haverhill resident Hannah Duston, along with her infant daughter, Martha, and her midwife, Mary Neff, were kidnapped in Haverhill by a Native American raiding party that intended to sell them as slaves to the French in Canada. This incident has been publicized for centuries and is much glorified, even though numerous Native Americans were slain.

The story goes that the group traveled by foot to an island in Boscawen, N.H., where the Contoocook and Merrimack rivers join. Duston and two fellow captives escaped after killing almost a dozen of their captors as they slept. Could this be true? Perhaps, and the legend persists.

Duston is considered to be the first American woman to have a statue erected in her honor. One stands in Haverhill's GAR Park, but there is another remembrance of her deeds near Boscawen. If Lowell and Lawrence were textile towns, Haverhill was known for its production of shoes. In 1832, there were twenty-eight shoe manufacturers in Haverhill.

A modern albeit bloodless story that makes one shake the head is the story of Bradford College. The academy opened in 1803 as one of New England's earliest coeducational

49

Above: New buildings are rising along the waterway.

Left: A statue of an armed Hannah Duston is one of Haverhill's most visible monuments.

institutions. In 1836, it began accepting only women. In 1932, it became Bradford Junior College, and in 1971, it began offering bachelor's degrees and admitting men again.

Yet the liberal arts institution closed in May 2000 because of financial problems. Its sudden demise surprised many North Shore residents, and it angered many students and their parents.

It seems like every community was scarred by fires that did enormous damage. The Great Fire of 1882 took place on February 17 of that year, which left shoe factories on Washington Street in ruins. It took only eight months for masons to rebuild and get the economy going again. The brick-and-stone buildings hummed with shoe-manufacturing and shoe retail businesses from the late 1880s to the mid-1900s. The trend today is to convert these former shoe-manufacturing buildings into residential units.

Almost every community on the Merrimack was damaged by the flood of 1936. That includes Haverhill. The snowfall in late 1935 and early 1936 was heavy and the cold lingered, leaving snow piled into March when several heavy rainstorms hit the northeast. The city's downtown was flooded, forcing evacuations and causing millions of dollars in damages to businesses and homes. School was canceled for four days. In the flood's aftermath, the Army Corps of Engineers built a 2,250-foot-long cement wall to keep the river from swelling into the city's downtown area again.

Northern Essex Community College opened in 1961 in the renovated Greenleaf Elementary School in Bradford. The college grew and in 1971 opened a spreading 106-acre campus near Kenoza Lake.

A fire of suspicious origin destroyed several buildings in the downtown Washington Street area along the Merrimack River in 1989. It was Haverhill's biggest fire in more than a century.

A massive Mother's Day flood in 2006 swept over the Merrimack Valley and southern New Hampshire, set off by heavy spring rains. The flooding affected communities along the Merrimack River, including Haverhill, where many homes and businesses along the river received heavy damage.

One of the key movers in the "Haverhill renaissance" has been Mayor James J. Fiorentini, who became mayor in 2003. A Haverhill native and a leading local lawyer, Fiorentini worked to improve schools, "greenify" the community and bring retail and hospitality trades back to the city. Many observers say his leadership is a key reason why Haverhill is enjoying a resurgence.[1]

Haverhill is mixing new construction with venerable infrastructure.

Cleanup specialist Rocky Morrison, shown here in stern of a vessel in Haverhill, has provided valuable service by removing debris from the river.

9

LAWRENCE:
A RIVERSIDE CITY DETERMINED TO IMPROVE

Lawrence is an example of a community that was created after the mills were sited and planned. To wit, Lowell had been so successful as a textile town that entrepreneurs went downriver to replicate their success in a town called Lawrence.

The Merrimack River is the reason that Lawrence exists. The city was planned to be a center of commerce. The river would provide the energy. Companies planned both for the manufacturing of products for profit and the housing and trade among employees that would create business away from the mills.

The Essex Company was a Boston-based corporation that organized the funding to create this industrial center. The sentiment was that if Lowell was a success (which it was), they would plan another such community. It was called Merrimack and later Lawrence (after impresario Abbott Lawrence). The Essex Company built a major dam in 1845, and incorporation of the town followed in 1847. It would employ thousands.

In the 1830s, Daniel Saunders, known as the founder of Lawrence, purchased strips of land on either side of the Merrimack River to gain control of waterpower rights. In 1843, he, along with other investors, formed the Merrimack Water Power Association and accelerated land purchases along the Merrimack, including a total of 7.5 sq. miles from Methuen and Andover, which would eventually become the city of Lawrence.

The Boston Associates had already developed nearby Lowell as one of the first planned industrial cities. They sought to replicate their success 10 miles downriver in Lawrence at the confluence of the Merrimack, Shawsheen, and Spicket rivers.

In so doing, key players—Daniel Saunders, Sr., John Nesmith, Edmund Bartlett, and Samuel Lawrence—took steps to secure capital and petitioned the state to build a dam and create and sell waterpower between Lowell and Lawrence. The influx of Boston capital created a mill city almost overnight, and for nearly 1 mile on both banks of the Merrimack, there stand the red brick walls of manufacturing.

The Essex Company built the industrial infrastructure and laid out streets, blocks of

53

Mills have lined the Merrimack for close to two centuries. This image is from the late nineteenth century. (*Lawrence History Center photo*)

The city of Lawrence is home to many scenic views of the river. (*Dyke Hendrickson photo*)

house lots, and parks. It imposed restricted use deeds—many still in force today—when selling lots or donating land to the new town. Restrictions included number, use, and location of structures on lots, height, and building materials.

The Essex Company, though its directors consisted of the interlocking Boston families that launched industrialization in New England, was led on the ground by Charles Storrow, the agent, chief engineer, and the city's first mayor. The records offer a glimpse of a man with a comprehensive vision and a determination to control its implementation.

Control in the creation of Lawrence meant not only state-of-the-art mills, but also corporation boarding houses on a scale large enough to enable mill owners to have sufficient sway over the behavior of their workers and to demonstrate to the world that workers could be accommodated in good-quality housing. It meant restricted deeds on lots to ensure that buildings were of sufficient quality. It meant micro-managing the development of churches, schools, and the local government.

The Boston Associates, with the creation of Lawrence, felt that they could not take a chance with the supply of water, and therefore, they created a company jointly owned by the Essex Company and Lowell's Proprietors of Locks and Canals to purchase all necessary land and water rights for the Merrimack up to and including Lake Winnipesaukee and the other large lakes of New Hampshire.

However, drinking water became a major issue in this community in the late nineteenth century. The city had a purification system, but it was not always effective. Indeed, typhoid afflicted many families along the river. Wealthy families lived away from the Merrimack and had private wells, but mill workers and their families got their drinking water from the river.

State public health officials were concerned about diseases being found in residents along the river. The Lawrence Experiment Station, now known as the Senator William X. Wall Experiment Station, was the world's first trial station for drinking water purification and sewage treatment. It was established by the Massachusetts State Board of Health. In 1886, the Massachusetts legislature required its Board of Health to adopt water pollution standards, which led to the creation of the station under the direction of Hiram Francis Mills, the "father of American Sanitary Engineering."

MIT professors William Ripley Nichols, Ellen Swallow Richards, and Thomas Messinger Drown were involved in some of the earliest research on sewage treatment and clean drinking water. At first, the station's main mission was to develop practical methods for treating wastewater. Its task was to determine the effect of filtration as compared to natural oxidation. During this time, scientists invented techniques for identifying and quantitatively analyzing the microorganisms in water and sewage. These studies helped set the standards in Massachusetts and, later, other states and countries.

In 1893, when a typhoid epidemic arose along the river, the city of Lawrence began filtration of river water using Mills' slow sand filters, thus becoming the first American city to filter its water for disease prevention. This filtering led to reductions in typhoid fever rate and overall death rate in the city.

The facility is now part of the Massachusetts Department of Environmental Protection (DEP), Division of Environmental Laboratory Sciences. It is responsible for providing technical and laboratory support to all DEP programs. In 2011, a $30-million, 13,000-sq. foot expansion was completed.[1]

The river is just a few feet deep here as it runs through parts of Lawrence. (*Dyke Hendrickson photo*)

Dams like this one in Lawrence help generate hydropower but make it impossible for fish to get upriver to spawn.

10

ELLEN SWALLOW RICHARDS WAS A PIONEER IN IMPROVING DRINKING WATER

The improvement of the Merrimack River has come as a result of political forces and a breakthrough in science. Two New England senators—Edmund Muskie (1914–1996; served in Senate 1959–1980) and George Mitchell (born 1933, served in Senate 1980–1996)—were effective in developing funding for sewage treatment solutions.

A scientist who was a pioneer in developing clean drinking water on the Merrimack was Ellen Henrietta Swallow Richards (1842–1911). She is recognized as leading a team to develop clean river water in Lawrence and Lowell. Her efforts stopped illness and death brought by drinking tainted water.

Richards was brought to the task when a state commission sought answers to numerous deaths of typhoid in about 1890. When typhoid rates spiked in the early 1890s, a team from Massachusetts Institute of Technology (MIT) was recruited to find solutions.

Many MIT men were involved in the project. Yet as a woman in a leadership position, Richards was so unique that people should know her name. For instance, in a poll of the greatest innovators in the history of MIT, she was ranked No. 8. That put her on the A-list of this writer.

Richards graduated from Westford Academy in 1862, and then came a lot of firsts. She was the first woman admitted to MIT. She graduated in 1873 and later became its first female instructor. Richards was the first woman in America accepted to any school of science and technology, and the first American woman to obtain a degree in chemistry, which she earned from Vassar College in 1870.

In 1873, she received a bachelor of science degree from MIT for her thesis, "Notes on Some Sulpharsenites and Sulphantimonites from Colorado." She continued her studies at MIT and would have been awarded an advanced degree, but MIT balked at granting this distinction to a woman. The university did not award its first advanced degree to a woman until 1886. Richards served on the board of trustees of Vassar College for many years and was granted an honorary doctor of science degree in 1910.

Ellen Swallow Richards was a pioneer in developing clean drinking water, which put an end to periodic outbreaks of typhoid. She was the first woman to graduate from MIT, in 1873.

In 1875, Swallow married Robert Richards (1844–1945), chairman of the Mine Engineering Department at MIT, with whom she had worked in the mineralogy laboratory. They took up residence in Jamaica Plain. With her husband's support, she remained associated with MIT, volunteering her services and contributing $1,000 annually to the "Woman's Laboratory." This was a program in which her students were mostly schoolteachers who wanted to perform chemical experiments and learn mineralogy.

Her first post-college career was as an unpaid chemistry lecturer at MIT from 1873 to 1878. It is remarkable to think an instructor was unpaid but she was a committed scientist. From 1884 until her death, Richards was an instructor at the newly founded laboratory of sanitary chemistry at the Lawrence Experiment Station on the river, the first in the United States.

Richards was a consulting chemist for the Massachusetts State Board of Health from 1872 to 1875, then the Commonwealth's official water analyst from 1887 until 1897. She also served as a nutrition expert for the U.S. Department of Agriculture.

She was interested in issues of sanitation, in particular air and water quality. At the request of the Massachusetts State Board of Health, she performed a series of water tests on 40,000 samples of local waters which served as drinking water for their immediate populations. These led to the so-called "Richards' Normal Chlorine Map," which was predictive of water pollution in Massachusetts.

As a result of her research, waters with chloride concentrations that deviated from the plot could be suspected of pollution. As a result, Massachusetts established the first water-quality standards in America.

Ever a multitasker, Richards also applied her scientific knowledge to the home. Since women were responsible for the home and family nutrition at the time, Richards felt all women should be educated in the sciences. Her book, *Food Materials and Their Adulterations* (1885) led to the passing of the first Pure Food and Drug Act in Massachusetts.

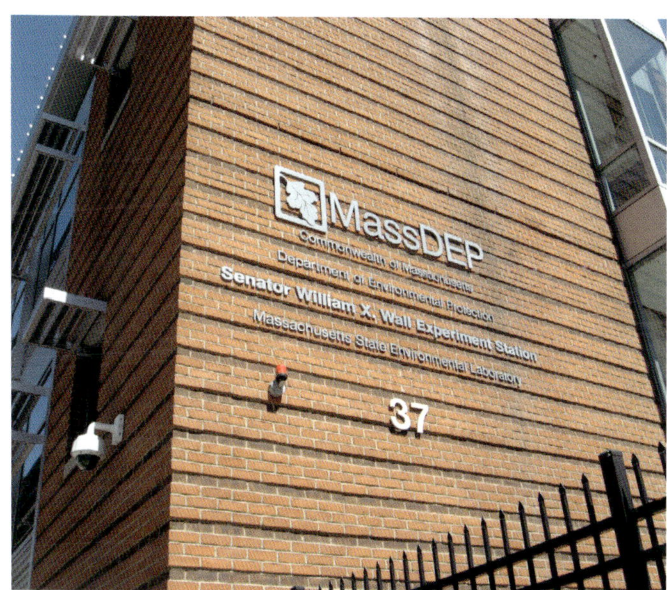

The Wall Experiment Station in Lawrence is a center for the study of clean drinking water.

She used her own home as an experimental laboratory for healthier living through science. Concerned with the air quality in her home, she moved from coal heating and cooking oil to gas. She and her husband installed fans to pull air from the home to the outside to create a cleaner air environment within the home. She also determined the water quality of the property's well through chemical testing, and to ensure that wastewater was not contaminating the drinking water.

Richards' interests also included applying scientific principles to domestic situations, such as nutrition, clothing, physical fitness, and sanitation. She is considered a "founder" of home economics.

Richards and Marion Talbot (Boston University class of 1880) became the "founding mothers" of what was to become the American Association of University Women (AAUW) when they invited fifteen other female college graduates to a meeting at Talbot's home in Boston, on November 28, 1881. The group envisioned an organization in which female college graduates would band together to open the doors of higher education to other women and to find wider opportunities for their training. The Association of Collegiate Alumnae (ACA), AAUW's predecessor organization, was officially founded on January 14, 1882.

She was also a leader in the drive to offer meals to in-need students. She was involved in an early program in some Boston high schools in 1894 to provide nutritional meals at low prices to children who would not normally have them. Due in large part to Ellen Richards and others, the New England Kitchen ran the program as a private enterprise that paid for itself many times over. It became the model for similar lunch programs in other cities.

Richards died in 1911 at her home in Jamaica Plain, after suffering from angina. Her work to ensure healthy drinking water still has merit today. More than 500,000 residents on the North Shore receive their drinking water from the Merrimack, which is a major reason by public officials should continue to monitor and upgrade the quality of the river.

11

LOWELL:
A VIBRANT HISTORY,
A REMARKABLE COMEBACK

The story of Lowell's development is closely connected to the Merrimack River. Once, it was nationally known as the largest textile manufacturer in the country. Now, it is famous for diversity. Dozens of nationalities live here, and UMass Lowell is one of the region's most successful centers of education. Like two centuries ago, leaders capitalized on technology to make Lowell a prosperous community.

Lowell has offered immigrants a chance. While some history texts stress the "exploitation" of mill workers and their families, the industry also provided jobs to many thousands of newcomers. Desperate families from Ireland, Greece, Poland, Armenia, Poland, Russia, and other countries found a place in Lowell.

Some families appreciated the mills. I have written two books about the French-Canadian migration to New England, and numerous "old-timers" said they valued the opportunity to work. In the nineteenth century, there were few jobs in Quebec. So, coming to the mills of New England provided an opportunity.

Many immigrants from Europe were more desperate. They fled starvation and even ethnic cleansing. The workers in Lowell, Lawrence, and other mill communities went on strike for better working conditions, but the mills also provided an opportunity. Perhaps one cannot generalize. Working in the mills suited some families and not others.

Patrick Jackson and others founded the Merrimack Manufacturing Company by opening a mill by Pawtucket Falls. They broke ground in 1822 and completed the first run of cotton in 1823. Within two years, a need for more mills and machinery became evident, and a series of new canals were dug, allowing for even more manufacturing plants.

With a growing population and booming economy, Lowell was spun off from Chelmsford. Named after Francis Cabot Lowell, it was officially chartered in 1826, with a population of 2,500. By the 1850s, Lowell had the largest industrial complex in the United States. Immigrants from around the world joined "Yankee mill girls" to work in the factories. Many French-Canadians came south for the winters and returned to work on their family farms in the summer.

The National Park Service gives tours of the Merrimack from its museum in Lowell.

Managers of Lowell mills were tight-fisted with money, as were most owners of the day. One of the less well-known facets of early mill life is that the "farm girls" that they employed were a feisty lot that sometimes protested low wages.

A strike took place in 1834. There was another brief strike in 1836 when room and board costs were increased. Work stoppages were not very successful against the corporation because workers were also boarders in corporate housing, and when they stopped work, there were no other jobs from which they could generate income.

Still, it is worth knowing that the women laboring along the Merrimack tried to improve their situations. There was a Factory Girls' Association formed in 1836. Then in 1845, the Lowell Female Reform Association was founded with about fifteen women. Yet women who were involved in labor movements were often blacklisted, and many left the mills to return to the family farms.

The textile industry wove cotton produced in the South. In 1860, there were more cotton spindles in Lowell than in all eleven states combined that would form the Confederacy.

The city continued to thrive as a major industrial center during the nineteenth century, attracting more migrant workers and immigrants to its mills. Later waves of immigrants came to work in Lowell and settled in ethnic neighborhoods. The city's population reached almost 50 percent foreign-born by 1900. It had more than 110,000 people in 1915.

When I wrote the book, *Quiet Presence: Stories of Franco-Americans in New England* (1980, Gannet Press, Portland), I talked to numerous workers in Lowell and Lawrence who were satisfied with their lives. Immigrants from Quebec especially saw good in the mill system. Many would work in the mills in the winter. With cash in pocket, they would return to the family farm in the summer. Some writers of the day say that mill cities in northern New England had "streets paved with gold" because cash wages were available.

The mill cities' manufacturing base declined as companies began to relocate to the South in the 1920s. The city fell onto hard times and was even referred to as a "depressed industrial desert" as only three of its major textile corporations remained active.

In recent decades, the city has begun to prosper again. UMass Lowell has provided a major economic boost. Numerous technology companies have settled in the area.

Lowell has a love of history. Its federal park in the downtown area is among the best in the country in telling the story of the mills. The facility offers tours on foot and by boat. Several mill weaving rooms have been recreated. Pedestrians will find that the city has numerous signs and placards that explain different elements of its past.[1]

Lowell hosts numerous canals, locks and dams that control the Merrimack to generate hydropower. The structure on the right is part of UMass Lowell.

The Merrimack is enjoyed by many residents in Lowell, for a variety of activities. (*Dyke Hendrickson photo*)

Lowell has utilized hydropower for two centuries and the city was a national leader in this technology.

12

MANCHESTER:
A THRIVING CITY IN NEW HAMPSHIRE

Manchester is a prosperous community. Its riverside mills, once among the largest in the world, have been filled with health firms, university classrooms, financial service centers, and restaurants. Other parts of the city host huge banks, hotels, and real estate offices. Every four years, the presidential primaries produce national media attention. Few mill communities in New England have prospered with the times as well as Manchester.

The city was named by the merchant and inventor Samuel Blodget. His vision was to create a great industrial center similar to that of the original Manchester in England, which was the world's first industrialized city. A century ago, the Amoskeag Company dominated the downtown. Thousands were employed by the mills, either directly or through secondary industries.

Travelers sometimes discuss the Merrimack in terms of "above Manchester"—ostensibly pristine—and below Manchester—the beginning of a string of industrial communities.

Native Pennacook Indians called Amoskeag Falls on the Merrimack River—the area that became the heart of Manchester—*Namaoskeag*, meaning "good fishing place." Yet the fishing did not stay good once the mills arrived.

In 1807, Samuel Blodget opened a canal and lock system to allow vessels passage around the falls, part of a network developed to link the area with Boston. He envisioned a great industrial center arising, "the Manchester of America," in reference to Manchester, England, then at the forefront of the Industrial Revolution.

In 1809, Benjamin Prichard and others built a water-powered cotton-spinning mill on the western bank of the Merrimack. Apparently following Blodgett's suggestion, Derryfield was renamed "Manchester" in 1810, the year the mill was incorporated as the Amoskeag Cotton and Woolen Manufacturing Company. It would be purchased in 1825 by entrepreneurs from Massachusetts, expanded to three mills in 1826, and then incorporated in 1831 as the Amoskeag Manufacturing Company.

Amoskeag engineers and architects planned a model company town on the eastern bank.

Manchester, N.H., was dominated for many years by the Amoskeag Manufacturing Co. In this photo, the mill was "dominated" by the Flood of 1936. (*Wikipedia photo*)

Incorporation as a city followed for Manchester in 1846, soon home to the largest cotton mill in the world—Mill No. 11, stretching 900 feet long by 103 feet wide and containing 4,000 looms.

Other products made in the community included shoes, cigars, and paper. The Amoskeag foundry made rifles, sewing machines, textile machinery, fire engines, and locomotives in a division. The rapid growth of the mills demanded a large influx of workers.

The Amoskeag Manufacturing Company went out of business in 1935, although its red-brick mills have been renovated for other uses. Few cities in New England have been able to repurpose the riverside brick structures as well as Manchester.

For many years, Manchester has been a major polluter of the Merrimack River. Yet good news came in the summer of 2020 when federal, state, and local officials announced a $231-million plan to improve their sewage-treatment operations.

Manchester officials, according to their public statements, want to do the right thing. Until this announcement, it seemed too expensive. Yet the twenty-year program will focus on cleaner effluent entering the Merrimack.

The Department of Justice and the federal Environmental Protection Agency were part of the agreement.

Because this development is potentially so powerful, the official statement is reprinted here. It might be boring and contain a lot of legalese, but because numerous government agencies are involved, the news is presented here.[1]

Manchester has developed a modern economy while filling many old mill buildings with colleges, tech companies, small businesses and restaurants. (*Dyke Hendrickson photo*)

13

MANCHESTER'S AGREEMENT TO PRODUCE CLEANER WATER

On July 13, 2020, the following news was announced to produce significant reductions in water pollution from Manchester:

The U.S. Environmental Protection Agency (EPA) and the U.S. Department of Justice (DOJ) announced an agreement with the City of Manchester that will result in significant reductions of sewage from the city's wastewater treatment systems into the Merrimack River and its tributaries. The State of New Hampshire joined the U.S. government as a co-plaintiff on this agreement, which also resolves alleged violations of the Clean Water Act by the City of Manchester.

Under a proposed consent decree filed in the U.S. District Court for the District of New Hampshire, the City of Manchester has agreed to implement a 20-year plan to control and significantly reduce overflows of its sewer system, which will improve the water quality of the Merrimack River. The plan is estimated to cost $231 million to implement.

The Merrimack River is a drinking water source for more than 500,000 people, is stocked with bass and trout for fishing, is used for kayaking, boating, and other recreational opportunities.

"This agreement means a healthier Merrimack River and cleaner water for the communities along the river in both New Hampshire and Massachusetts," said EPA New England Regional Administrator Dennis Deziel. "EPA has long been committed to working with our state and federal partners and cities like Manchester to improve water quality along the Merrimack, which is an important source of drinking water and recreation destination."

The settlement addresses problems with Manchester's combined sewer system, which when overwhelmed by rain and stormwater, frequently discharges raw sewage, industrial waste, nitrogen, phosphorus, and polluted stormwater into the Merrimack River and its tributaries. The volume of combined sewage that overflows from Manchester's combined

Manchester, like other communities on the Merrimack, has benefited from the river's falls.

sewer system is approximately 280 million gallons annually, which is approximately half of the combined sewage discharge volume from all communities to the Merrimack River.

Under the proposed consent decree, Manchester will implement combined sewer overflow (CSO) abatement controls and upgrades at its wastewater treatment facilities that are expected to reduce the city's total annual combined sewer discharge volume by approximately 74 percent from approximately many million gallons annually to 73 million gallons.

The two major components of the CSO abatement controls will disconnect Cemetery Brook in Manchester, the largest of the local five significant connected brooks, from the city's combined sewer system. Manchester will design and construct a new 2.5-mile drain for Cemetery Brook from Mammoth Road to the Merrimack River to convey both the brook's and storm drainage flows. The city will also design and construct projects to separate the combined sewers for areas adjacent to the Cemetery Brook drain. These drainage and sewer separation projects will together address the largest drainage basin in the city and produce the greatest volume of CSO reduction.

The work under the proposed consent decree also includes the construction of a new drain and sewer separation in the Christian Brook drainage basin, which will remove the third-largest brook from the wastewater collection system.

The river runs rapidly in the spring through Manchester.

Here the Nashua River meets the Merrimack in Nashua, N.H. (*Dyke Hendrickson photo*)

The proposed consent decree also requires the city to implement a CSO discharge monitoring and notification program, which will include direct measurement of all discharges from six CSO outfalls estimated to be more than 99 percent of all of the city's total CSO discharge volumes.

The city will be required to provide initial and supplemental notification to the public, including public health departments and downstream communities, with notification made through electronic means such as posting to the city's publicly available website and reasonable efforts to provide other notification.

In addition to the twenty-year control plan, the proposed settlement also requires the upgrades to improve the handling of solid waste at the wastewater treatment plant to reduce discharges of phosphorus.

Many of the communities in the Merrimack River watershed are environmental justice communities with large numbers of minority and low-income residents.

In September 2019, the EPA issued Clean Water Act permits to the cities of Haverhill, Lawrence and Lowell, Massachusetts, under the National Pollutant Discharges Elimination System to reduce pollutant discharges from the three wastewater treatment plants and associated CSOs in the Merrimack River at 27 locations across the three cities.

14

FRANKLIN, N.H.:
TWO RIVERS RUN THROUGH IT TO FORM THE MERRIMACK

Those visiting Franklin, N.H., do not see many signs that it represents the start of the Merrimack River. One of its unique characteristics, though, is that it has plaques and memorials that remember the paper mills that operated there.

Most Merrimack communities hosted textile mills. Yet Franklin is so far north that its history is in lumber and papermaking.

The Merrimack begins where the Pemigewasset River merges with the Winnipesaukee River. That exact point is somewhere behind Franklin High School, but as has been stated, the actual point of the merger has not been glorified.

The Pemigewasset originates at Profile Lake in Franconia Notch State Park, in the town of Franconia. It flows south through the White Mountains and merges with the Winnipesaukee River to form the Merrimack in Franklin. The Merrimack then flows through southern New Hampshire, northeastern Massachusetts, and into the Atlantic.

The Pemigewasset watershed consists of over 1,100 miles (1,800 km) of rivers and 17,000 acres (69 km^2) of lake, pond, and reservoir area. The watershed comprises about 20 percent of the Merrimack's total watershed area.

The Winnipesaukee River is 10.5 miles long. The river's drainage area is approximately 488 sq. miles. Some adventurers, like the Voyagers, begin in Franklin so they can complete the full 117 miles. Others, like the adventurer in our next segment, start in Manchester.

Left: The Merrimack begins in Franklin, N.H.

Below: Some travelers start their trip downriver from Franklin.

15

TRAVEL ESSAY:
AN ADVENTURER TAKES A TRIP ON THE MERRIMACK

The Merrimack River has always generated interest. Here are excerpts from a "travel piece" written in 2003 and appearing in *American Heritage* magazine. The author is Rosanne Haggerty, an amateur adventurer and MacArthur Fellow scholar. She now heads an organization to develop affordable housing in New York City. She writes about a trip taken down the Merrimack. It is included here because she deftly combines history, critical assessments, and travel writing to explain her voyage on the Merrimack.

Matters did not look promising. the path down to the canoe launch onto the Merrimack River was long and steep, thick with roots and brambles and sharply angled. Pushing, pulling, and grunting, we reached a scum-slicked spit of sand just below a wide stretch of renovated nineteenth-century mill buildings in Manchester, New Hampshire, and pushed off.

In other words, the Merrimack does not always make it easy for boaters to play on its surface. But that is not surprising; until quite recently no one would have wanted to. Its image problems go back more than a century. Described by National Geographic in 1951 as "a veritable slave in the service of industry," this 117-mile-long river was dammed, canalled, and dumped on to within an inch of its life. Longtime residents recall watching what they sometimes called the "Merrimuck" change color depending on which dyes the textile mills were using that day. Its vegetation grew in mutant forms. As a repository for everything from medical waste to offal, it reeked.

Why bother with the Merrimack? Because it encapsulates much of New England's history, colonial, industrial, and post-industrial. Because Henry David Thoreau traveled along it to write his elegiac *A Week on the Concord and Merrimack Rivers*. And because it is beautiful.

The waters of the Merrimack powered America's first Industrial Revolution, initially for the textile mills of Lowell, Massachusetts, and then for those in communities like

The Merrimack River got cleaner after passage of the Clean Water Act of 1972.

Manchester and Nashua, New Hampshire, and Lawrence and Haverhill, Massachusetts. But just as the river was sacrificed with little regard for its long-term health, so the vast mill complexes died from underinvestment and resistance to change.

The river's comeback began with the Clean Water Act of 1972, which required that sewage be treated before it reached the nation's waterways.

And not just people appreciate the difference. Bald eagles now nest on the Merrimack's banks, as do ospreys, egrets, and hawks. American shad, striped bass, trout, and Atlantic salmon swim in its depths. Otters and minks frolic on its shores. What happened was an unusual symbiosis. Communities acted to regenerate the river, an effort that today embraces hundreds of volunteers who monitor its temperature, pulse, and respiration, and the river returned the favor, helping regenerate the communities on its shores.

The Merrimack officially begins where the Winnipesaukee and Pemigewasset rivers join in Franklin, New Hampshire, the birthplace of Daniel Webster, whose home you can visit. Determined to start at the exact beginning of the river and assured that we couldn't miss it (which naturally made us nervous), we parked downtown and asked around for the source of the Merrimack. We might as well have been asking for the source of the Nile. Before long, though, we think we have found its origins behind a parking lot.

For a body of water once so famously tainted, the stretch from Franklin to Concord is improbably bucolic. Mile after mile there is little to see but birdlife, trees, fields of corn, and the occasional church steeple poking up above the vegetation. The Pennacook Indians, who were there long before the factory builders came, might well find the scene familiar.

We ended our day of paddling in Boscawen, just north of Concord, where perhaps the most politically incorrect monument in America bears witness to the difficult relations between Native Americans and early European settlers. A short walk away from U.S. Route 4, it is a 35-foot-high granite statue to Hannah Duston, erected in 1874 and the first statue in America dedicated to a woman. Duston's story began 60 miles away in Haverhill, Massachusetts, which has a more elaborate monument to her. There, on the morning of March 15, 1697, Indians descended on the town, burned half a dozen buildings, and took several people captive, including Dustin, then recovering from the birth of a child. The baby was killed in the raid. On March 31, held on an island in the Merrimack in what is now Boscawen, Hannah and two companions stole some hatchets and struck back, killing 10 of their 12 captors. They then coolly took the scalps, made their way home, and collected a bounty. Boscawen's statue to Hannah commemorates the spot where she killed the Indians. Stepping forward with a tomahawk in her right hand and a clutch of scalps in her left, she is clearly a woman to be reckoned with.

The river runs through Concord, New Hampshire, but somehow the state capital is not a river town. Concord sits back from the water, and unlike every other city on the route, it lacks massive squat brick mills, favoring instead graceful Victorian architecture and granite buildings designed to impress people with the seriousness of government.

The first of the great Merrimack mill towns is Manchester, about 20 miles to the south. On a sunny Sunday in August, the water is so clear we can read the labels on the refrigerators and tires dumped in the past and not yet cleared away; the air is pure, the scene quiet. Even as we skim past a full mile of mill buildings, it just doesn't seem possible that for a century this was a pulsing center of industry.

Manchester grew up around the Amoskeag Mills, which grew up around the Amoskeag Falls of the Merrimack River. In its day, the Amoskeag mill complex was the world's largest, employing 17,000 people in the early 1900s. In their uniformity and scale, Amoskeag's buildings resembled a medieval walled city, an all-inclusive social.

By the early 1920s, the currents that were to undermine all of New England's nineteenth-century textile production—failure to modernize, labor problems, competition from the South, and excess capacity—had weakened the Amoskeag. In 1936, after a string of punishing strikes, it closed, devastating the local economy. Merchants made an effort to attract small-scale businesses into the abandoned mills, but many of the buildings didn't see a tenant for 40 years, and most of the businesses that did move in were gone by the mid-1970s.

Following more familiar economic development strategies, the CPR created a downtown master plan and a business improvement district and built a civic center. The city has attracted educational institutions like New Hampshire College to the mill yard, established a historic district and redeveloped a former military base into Manchester Airport. Manchester has clearly come back from its dark days. The downtown is active with restaurants, shops, and cultural institutions, and the mill yards have probably never looked better.

The best white water on the river surges between Manchester and Merrimack, which are also rich in remains of the Middlesex Canal, built in 1803, an engineering warm-up for the Erie Canal. The Middlesex was built to bypass the rapids and allow for clear passage

In communities like Methuen, the river offers soothing views for boaters and passersby alike.

Magnificent views can be had at Moseley Woods Park in Newburyport.

The clock above the mills is an iconic feature of Lawrence.

of the lumber from northern New England, used by the shipbuilders of Newburyport. Later, during the factory-building era, Merrimack Valley farmers discovered a lucrative sideline making bricks from local clay. But the mills used the bricks to build dams to spin turbines to drive textile looms, and the river's purpose became power, not transportation. Goodbye, canal-boats; hello, railroads.

Lowell's modern history began in the 1820s. A hard-eyed business venture from the start, it was nevertheless informed with a visionary idea—that the miseries of industrialism in Europe need not be replicated in America. A group of merchants known as the Boston Associates set about using the Pawtucket Falls, where the Merrimack drops 32 feet, to power a factory complex that, although enormous, would create no permanent working class. The mill hands would be local farm girls who would return to their homes after a few years of wage-earning and self-improvement. Good working conditions, albeit for 14 hours a day, a church, supervised boarding homes, a program of lectures and cultural enrichment—all would prove that America could have industrialization without the horrific social effects that came with it in England.

In 1826, Lowell had just 2,500 people and a few looms; by 1850 there were 35,000 residents, and 10,000 workers were producing almost 2 million square feet of cloth a week. Labor relations were not as smooth as the official story claimed. The "mill girls" struck three times between 1830 and 1840, protesting wages and working conditions and demanding a 10-hour workday, which finally came—for women and children—in 1874. After the Civil War, perhaps weary of dealing with these tiresomely independent Yankee females, the mills' managers turned to immigrants, particularly to the Irish and French Canadians. These newcomers were willing to work longer hours for less money, and they neither required nor wanted paternalistic supervision. The transition to immigrant labor made economic sense, but it also spelled the end of the Boston Associates' grand experiment and the beginning of the industrial working class they had been determined to never let form.

Lowell prospered through the First World War, its population peaking in 1920 at just under 113,000. But then, as in Manchester, the mills began to decline. By the mid-1950s the last of the original ones had closed.

Much of the credit for the city's turnaround goes to local leaders like the late Sen. Paul Tsongas, a native of the town, U.S. Rep. Brad Morse, and the educator Patrick Moogan, who imagined that the story of Lowell's beginnings as the nation's first planned industrial city could be the key to its renewal. In contrast to the individual entrepreneurship and locally funded efforts that returned Manchester to its feet, Lowell relied on massive federal and state investments. The big break came in 1978 when Congress designated it the nation's first urban National Historical Park.

A few miles downriver the Merrimack takes a sharp northeastern turn and passes through small towns and suburbs on its way to another remnant of the Industrial Revolution. This is Lawrence, which is a kind of anti-Lowell, the image of what Lowell would have been without the national park, Paul Tsongas, and all the rest. Clearly, the comparison hurts.

Its dam was the biggest in the world. The clock tower that looms over the city is the world's second-largest, just 6 inches smaller than Big Ben. And the 1912 "bread and roses" strike was a turning point in labor history.

Lawrence is impressive. Its old mills line the river as far as the eye can see, and they trace the early designs of its planners. Boardinghouses for workers were sited immediately parallel to the mills (one of which has been restored as the Lawrence Heritage State Park), followed by a row of commercial and municipal buildings and then another of houses of the gentry. Lawrence is a work in progress. Still, it is progress, fueled by a population that, as before, is composed largely of immigrants, this time from Latin America.

One of the intriguing facts of life in New England is how towns next door to each other can have such different characters. Haverhill lies on the same side of the Merrimack as Lawrence and just a few miles away, but it's a completely different kind of place. Though for some years it was a major shoemaking center, the "Queen Shoe City of the World," Haverhill looks and feels like the old New England village it still is.

But it is Newburyport, where the Merrimack opens into the Atlantic, that beckons us on this stretch of the river. Newburyport was never as rough-edged as Manchester, Lowell, Lawrence, or even Haverhill. Still, being downriver from all those factories took its toll.

Yet Newburyport's history is entirely different. European settlers founded it in 1635, quickly built a church, and then proceeded to engage in three decades of arcane doctrinal disputes.

Diverting as the religious disputes were, by the end of Newburyport's first century much of its attention had been given over to shipbuilding. The region's pine and oak provided raw materials, and the town's access to the Atlantic made it an ideal location. For a hundred years, Newburyport was one of the nation's leading shipbuilders, turning out thousands of vessels. The late 1700s and early 1800s were Newburyport's glory days, as it grew rich on the ships it built and the cargo they brought home, a good deal of it illicitly.

A waterfowl loiters in Lawrence. Parts of the river in this community support much wildlife.

Small vessels tied up near Cashman Park, Newburyport. In background is the marsh, which is the largest in New England.

Then protectionist legislation in 1807 and 1812 killed off international trade, while the Middlesex Canal diverted traffic to Boston; in 1811 a fire swept through downtown. But the town rebuilt itself, this time in brick, and endured as an important point of entry for foreign vessels. The mortal blow came in the 1840s when railroads displaced the port. Like many of its upstart neighbors, Newburyport turned to cotton mills and immigrants for its economic survival. When the mills went overseas, or south, Newburyport hit the skids.

Newburyport never lost its spirit. In the early 1960s, when it had reached its nadir of dilapidation, with much of the downtown boarded up, the town officials took action. A band of saviors from the morning time of the preservation movement stepped in. They formed a committee and decided it would pursue preservation, not box stores.

16

History of a Cleaner Merrimack River:
Creation of EPA Led to Improvement

In order to understand the improvement in water quality of the Merrimack—and the subsequent improvement of life along the river—it is necessary to recognize the work of national leaders about fifty years ago.

The Environmental Protection Agency was founded in 1970, formed during the administration of President Richard Nixon. However, two senators from New England—Edmund Muskie, D-Maine, and George Mitchell, D-Maine—played key roles in cleaning American rivers.

Muskie, who served 1959–1980, grew up on the polluted Androscoggin River in Rumford. Mitchell, who served in the Senate from 1980–1996, was raised on the troubled Kennebec River in Waterville. Muskie was responsible for the passing of the Clean Water Act of 1972. Mitchell played a key role in blocking President Ronald Reagan's plan to abolish funding. Mitchell's leadership resulted in the passage of the Clean Water Act Amendments of 1987.

As activists in 2020 sought to keep the Merrimack Water clean, the words and actions of these senators provide lessons on commitment and political skill to obtain crucial funding to help clean rivers.

In both 1972 and 1987, a Republican president occupied the White House. They were not advocating for clean water. Yet these two Democratic senators—by generating bi-partisan support—provided the leadership that led to the passage of bills that cleaned many rivers, lakes, and oceans. Indeed, it appears that constant bi-partisan action by officials and residents of the North Shore is necessary to keep the importance of a clean Merrimack on the local political docket.

Federal officials are essential, but local leaders, too, must lobby for a clean river. Communities including Lowell, Lawrence, Haverhill, and Manchester, N.H., must spend more to improve their sewage-treatment plants. Every riverside community can cut down on the amount of debris and dirty run-off that goes from the streets to the river. They can make financial commitments to upgrade their sewage-treatment plants.

Sen. Edmund Muskie, D-Maine, was the key figure in leading to the passage of the Clean Water Act of 1972. Seen in the background is his wife, Jane.

The EPA, with its regional office in Boston, is a major force in the health of the Merrimack River. Though some environmentalists on the North Shore show impatience with local EPA officials, many elected leaders feel that the EPA is an essential force working to improve the Merrimack. In 2018, the national EPA had 14,172 full-time employees. More than half of EPA's employees are engineers, scientists, and environmental protection specialists; other employees include legal, public affairs, financial, and information technologists.

Many public health and environmental groups advocate for the agency. The Trump Administration cut the EPA budget and nominated some unlikely leaders, but the EPA is still a force in supervising New England rivers.

Looking back, it is difficult to imagine that not until the 1950s did a mass of Americans become concerned about pollution in rivers, lakes, and oceans. Many older North Shore residents remember the Merrimack as filthy. Children were forbidden to get near it. In an earlier epoch, it was capable of spreading diseases like typhoid. In the late 1950s and through the 1960s, Congress reacted to increasing public concern about the impact that human activity could have on the environment. In 1959, Congress passed the Resources and Conservation Act that proved to be an important initial step.

The 1962 publication of Rachel Carson's *Silent Spring* alerted the public to the damaging effects of pesticides. Film clips from that era show her as a brave advocate testifying in Congressional hearings. Many conservative legislators mocked her findings. In the following years, federal bills were introduced, and hearings were held to discuss the environment.

One infamous event that helped spark the environmental movement was the burning

The drawbridge in Newburyport often is part of scenic sunsets. (*Dan Graovac photo*)

of Ohio's Cuyahoga River in 1969. The bizarre nature of the story and graphic photos led to a national outcry. In December 1970, a federal grand jury investigated water pollution allegedly being caused by twelve companies in northeastern Ohio. A bill was introduced in Congress to create an agency that could help clean filthy waterways.

President Richard Nixon signed the National Environmental Policy Act into law in January 1970. The measure created the Council on Environmental Quality in the executive office of the president. NEPA required that a detailed statement of environmental impacts be prepared for all major actions significantly affecting the environment. The "detailed statement" would ultimately be referred to as an environmental impact statement. In July 1970, Nixon proposed an executive reorganization that consolidated many environmental responsibilities of the federal government under one agency—the new Environmental Protection Agency. After conducting hearings during that summer, the House and Senate approved the proposal.

The EPA was created ninety days before it had to operate, officially opening its doors in December 1970. The agency's first administrator was William Ruckelshaus. In its initial year, the EPA had a budget of $1.4 billion and 5,800 employees. At its start, the EPA was primarily a technical assistance agency that set goals and standards. Soon, new acts and amendments passed by Congress gave the agency its regulatory authority.

When EPA began operations, members of the private sector felt that the environmental protection movement was a passing fad. Ruckelshaus stated that he felt pressure to show the public that EPA could respond effectively to widespread concerns about pollution.

Looking into the fire on the Cuyahoga River was the first grand jury investigation of water pollution in the area. The attorney general of the United States, John N. Mitchell, held a press conference referencing new pollution control litigation, with a focus on work with the new Environmental Protection Agency. He announced the filing of a lawsuit against the Jones and Laughlin Steel Corp. for discharging substantial quantities of cyanide into the Cuyahoga River near Cleveland.

It sounds counterintuitive today to suggest that Richard Nixon and John Mitchell, two rogues of the Watergate scandal, were leaders of the early environmental movement, but it began during the Nixon Administration, and their names are on the documents. Also, Democrats were energized to work for more legislation. Congress enacted the Federal Water Pollution Control Act of 1972, better known as the Clean Water Act. This measure established a national framework for addressing water quality to be implemented by the agency in partnership with the states. It also made available billions of dollars in grants for communities to build sewage-treatment plants.

In 2020, there are few free grants to be had. Local representatives are attempting to access low-interest loans to improve sewage-treatment facilities. It is important to note that a loan is different from a grant. The grants of the 1970s came with minimal financial obligation for communities that said they lacked the money to do it themselves. Many communities in 2020 find it difficult to gain local support for spending for clean water.

Yet here, the EPA is helpful. If the federal agency takes action against a municipality or regional water district, a community can be compelled to generate the money. In 2020, Manchester, N.H., agreed to a $232 million improvement program that will take twenty years to complete. It did so because federal and state authorities were compelling them to do so. The era of EPA monitoring and clean-up is upon us, and the Merrimack River is a beneficiary.

Where the Merrimack meets the Atlantic. Note the sandbar, which has frustrated mariners for many years.

17

SPECIFICS OF THE CLEAN WATER ACT

The Federal Water Pollution Control Act of 1948 was the first major, if weak, U.S. law to address water pollution. Growing public awareness and concern for controlling water pollution led to sweeping change in 1972. As amended in 1972, the law became commonly known as the Clean Water Act (CWA), which did the following:

- Established the basic structure for regulating pollutant discharges into the waters of the United States.
- Gave EPA the authority to implement pollution control programs such as setting wastewater standards for industry.
- Maintained existing requirements to set water quality standards for all contaminants in surface waters.
- Made it unlawful for any person to discharge any pollutant from a point source into navigable waters, unless a permit was obtained under its provisions.
- Funded the construction of sewage treatment plants under the construction grants program.
- Recognized the need for planning to address the critical problems posed by nonpoint source pollution.

Subsequent amendments modified some of the earlier Clean Water Act provisions. Revisions in 1981 streamlined the municipal construction grants process, improving the capabilities of treatment plants built under the program. Changes in 1987 phased out the construction grants program, replacing it with the State Water Pollution Control Revolving Fund, more commonly known as the Clean Water State Revolving Fund. This new funding strategy addressed water quality needs by building on EPA-state partnerships. Yet they were loans, not grants. Congressional representative Lori Trahan, D-Lowell, was a leader among North Shore leaders in 2020 attempting to access the State Revolving Fund.

The story of Hannah Duston is still told along the river. Sometimes, as above, her name is spelled as Dustin.

18

Muskie's Clean Water Act Saved Merrimack and Other Rivers

Sen. Edmund S. Muskie will be remembered as a presidential candidate (1972) and a strong Democratic Senate leader, but those who value rivers will recall his groundbreaking work leading to the passage of the Clean Water Act of 1972. The formal name is the Federal Water Pollution Control Act Amendments of 1972. Though Muskie was a Mainer ("I'm not from Maine, I am of Maine," he used to say), the work of the persistent senator helped scores of rivers across the country, including the Merrimack.

The Clean Water Act outlined what responsibilities cities had in developing sewage-treatment plants and limiting the entry of raw sewage into waterways, and it provided grants, not loans, to get the program started. Though the Nixon Administration helped create the Environmental Protection Act, Nixon vetoed the Clean Water Act. Yet the measure was passed over the veto. Clearly, the bill's time had arrived and much of the reason it was put into law was because of Ed Muskie and his ability to develop bipartisan support.

Muskie knew pollution. I was a young reporter with the Portland Press Herald who traveled with him and several aides in 1976 during his re-election campaign for Senate against Republican Robert Monks. We went to a small factory near South Paris, Maine. The factory, which employed less than a dozen, was remarkably dirty and ill-kempt. In addition, odors from paint and turpentine suggested that the lighting of a match would turn the place into an inferno. Muskie put a handkerchief up to his nose, but he took the whole tour. His influence could have led to its shutdown, but the senator knew when to choose his battles.

After a tour with the owners, Muskie left. Everyone in our car was coughing, and we were laughing. Perhaps we were high from inhaling the fumes. It was the filthiest stop on our tour, but it was clear that Muskie did not force a confrontation with every company owner. Yet when he did enter a fight against pollution, he was formidable.

This essay was produced by the Maine Historical Society, Maine Memory Network:

Muskie grew up in Rumford near the Androscoggin River, which received pollutants from paper mills and municipal sewage as well as sources like agricultural and street run-off. The geographic and geologic attributes of the Androscoggin, like other rivers in New England, made it ideal for industry—and pollution. The Androscoggin River, from its source in the Rangeley Lakes to its outlet in Merrymeeting Bay, drops 1,500 vertical feet. Those drops are excellent sources of waterpower and drew industry to the river's shores beginning in the early nineteenth century.

Before Muskie went to the U.S. Senate in 1959, the problems of the Androscoggin River had come to the fore through the work of Dr. Walter Lawrance, a Bates College chemistry professor. Since the 1880s when paper mills began creating pulp with a sulfite chemical process, water downstream of paper mills often smelled awful (like rotten cabbage) and was too polluted to safely drink. A study of the Androscoggin in the early 1940s showed that it had almost no dissolved oxygen and therefore could not support fish or aquatic life.

Lawrance introduced sodium nitrate into the river to improve its oxygen levels and hence its odor and the health of the river. While Walter Lawrance worked to improve the water quality of the Androscoggin River, Muskie operated on a larger stage. When he was elected to the U.S. Senate in 1958, Muskie was assigned to the Public Works Committee, which had little influence. Now known as the father of the modern environmental movement, Muskie turned the committee and its Subcommittee on Air and Water Pollution, formed in 1965, into important voices.

Muskie believed that environmental regulations needed enough science and scientific analysis to justify the Federal government getting involved. Muskie and the committee wanted the Public Health Service, which had a charge to deal with pollution, to collect scientific evidence of the effects of pollution on health and welfare.

Leon G. Billings, the former staff director of the Air and Water Pollution subcommittee, wrote, "Better science argues for more, rather than less, aggressive pollution controls." Muskie was known for thoroughly studying issues and basing his arguments for legislation on the facts he gathered.

Muskie shepherded the 1963 Clean Air Act through Congress and stood close to President Lyndon Johnson as he signed the landmark law. Muskie later said his greatest achievement was the 1970 Clean Air Act, an act that led the way to the 1972 clean water bill. As a number of Muskie's assistants and colleagues from the 1960s note, "environmental" was not a term much used then. There was a conservation movement that focused on preserving wild lands. Some wanted better fishing and hunting. Muskie's focus was different—he wanted to clean up the air, water, and land for everyone.

When the polluted Cuyahoga River in Cleveland caught fire in 1969, Muskie's crusade for clean water drew a larger audience. Muskie's reputation as a steward of the environment earned him the nickname "Mr. Clean" and drew letters from people around the country seeking his intervention or support for clean air and clean water causes.

Suzanne Clune, an 11-year-old from Poland, Maine, wrote to Muskie about the odor of the Little Androscoggin River near her home. She said area residents had written to the governor with no result and urged Muskie to help the situation. A Portland dentist wrote to Muskie, enclosing photographs of pollution of Casco Bay at the outlet of the Presumpscot River and asking Muskie to take action.

A dentist, Richard E. Gosse, wanted industries, especially the paper mill upriver in Westbrook, held accountable for river pollution. Gosse was just one of hundreds of people who wrote to Muskie, expressing their concerns and helping to strengthen his case that the federal government must do something to protect the environment.

Muskie's popularity and position as an environmental leader gave him the opportunity to speak about the challenges of dealing with pollution and other problems. He called dealing with pollution a "highly difficult and potentially explosive political opportunity," and noted there were hard choices to be made, especially at the local level.

Muskie introduced what came to be known as the Clean Water Act in October 1971. His Senate subcommittee held 33 days of hearings over a period of two years in various locations around the country before and after the bill was introduced.

The legislation that Muskie and the strongly bipartisan committee backed was bold. It required zero discharge by 1985. The goal was that nothing of a polluting nature ultimately would be allowed to be dumped into rivers and lakes directly and indirect run-off would be controlled as well.

Previously, water pollution was discussed in terms of how much discharge from various sources the body of water could stand. Muskie and the committee believed a goal of no pollution was the best alternative to trying to decide what and how much of various materials would be acceptable in a body of water.

It gave the Environmental Protection Agency the authority to implement pollution control programs that promoted "biological integrity," a term that became especially important to continued environmental efforts. Adding to the disputes over discharges, cost, and enforcement were political tensions arising from the 1972 Presidential election.

Some residents think the beauty of the Merrimack contributes to their mental health.

Both Ed Muskie, a Democrat, and Richard Nixon, the Republican incumbent, were running for president. Even though Muskie dropped out of the race, accusations of political posturing continued.

Nixon was not the only opponent of the bill and the standards it created. Some environmentalists opposed it as well because they did not trust the experts that the committee had relied on in constructing the bill. They also did not trust the government enforcement agencies such as the Environmental Protection Agency and the Army Corps of Engineers.

Many people contributed to writing and marking up the bill, and Muskie worked with colleagues of both political parties to settle on language and provisions before sending the legislation to the floor of the Senate for a vote. It passed 86-0 in the Senate and 380-14 in the House. The House and Senate then spent thirty-nine days in conference to work out their differences. The Conference Committee report was passed with similar large margins.

President Richard Nixon vetoed the 1972 Clean Water Act, but on October 18, 1972, the veto was overridden and the bill became law. While the no-discharge goal that Muskie and his allies promoted has yet to be achieved, the 1972 Clean Water Act is credited with leading to improved water quality in rivers, lakes, and streams. In the years since the bill passed, agricultural and urban run-off, in particular, have been greatly reduced.

19

SEN. GEORGE MITCHELL DEVELOPED THE CLEAN WATER AMENDMENTS OF 1987

In the late 1980s, President Ronald Reagan wanted to cease giving grants that financed sewage treatment plants and other clean-water initiatives. Yet financial support of sewage-treatment programs was important to communities tied to lakes, rivers, and streams, like towns and cities on the Merrimack. Democrats rose up in opposition.

Sen. George Mitchell, a Democrat from Maine like Muskie, led the drive to maintain funding for clean water. He was able to keep the program alive, albeit in a weaker version. The amendments to the Clean Water Act of 1972 were passed in 1987. One key provision was that they provided low-interest loans—not grants—to communities that sought financial support.

Reagan was successful in abolishing outright grants, but clean-water advocates, led by Mitchell, passed legislation that provided for low-interest loans. It is still known as the Clean Water State Revolving Fund.

According to the EPA, this is a powerful partnership between EPA and the states that replaced EPA's Construction Grants program. States have the flexibility to fund a range of projects that "address their highest priority water quality needs." The interest rate in 2019 was 1.5 percent, according to federal figures.

Using a combination of federal and state funds, CWSRF programs provide loans to achieve the following: construct municipal wastewater facilities; control nonpoint sources of pollution; build decentralized wastewater treatment systems; create green infrastructure projects; develop project estuaries; and fund other water quality projects.

The program has provided $138 billion to communities through 2019, federal officials say. States have provided 41,234 low-interest loans to protect public health, protect valuable aquatic resources, and meet environmental standards benefiting hundreds of millions of people.

The story of cleaning the Merrimack is as much a tale of politics as it is science. The following are a transcription of minutes from a federal hearing almost two decades ago.

Former Sen. George Mitchell, D-Maine, fought to save the Clean Water Act by leading legislators to pass valuable amendments in 1987.

It features an address by Mitchell, who retired in 1996, contemplating the passing of this act in 1987. It is relevant because it also references Muskie.

The work of Mitchell, Muskie, and other clean-water legislators is a major reason why the Merrimack became cleaner after 1972. In 2020, leading elected officials in Massachusetts were seeking funds from the Clean Water State Revolving Fund.

COMMEMORATION OF THE THIRTIETH ANNIVERSARY OF THE CLEAN WATER ACT

Hearing before the Committee on Environment and Public Works, U.S. Senate

Oct. 8, 2002
U.S. Senate, Committee on Environment and Public Works, Washington, D.C. The committee met, pursuant to notice, in room 406, Senate Dirksen Building, Hon. James Jeffords (chairman of the committee) presiding. Present: Senators Jeffords, Bond, Carper, Chafee, Clinton, Voinovich and Wyden.

Statement of the Hon. George Mitchell, a retired U.S. Senator from the state of Maine.

Senator Mitchell: "Thank you very much, Mr. Chairman, and members of the committee. I appreciate the opportunity to join you today on the 30th anniversary of the passage of the Clean Water Act, especially in the company of my friend and colleague, Senator Stafford.

"We have made progress since 1972 in meeting the goal of the Act, which is, as Senator Voinovich noted, to restore and maintain the chemical, physical and biological integrity of the Nation's waters. Our nation has invested nearly $75 billion to construct municipal sewage treatment facilities, nearly doubling the number of people served with secondary treatment to almost 150 million. However, there is much more to be done.

"The EPA's Assistant Administrator for Water said recently that about 40 percent of our Nation's waters do not meet fishable, swimmable standards. That bears repeating. After 30 years of implementing the Clean Water Act, 40 percent of our Nation's waters remain impaired. Clearly, we must intensify our efforts.

"I would like first, Mr. Chairman, to recognize the contribution of one of our Nation's great pioneers in environmental legislation, my friend and mentor, Senator Edmund Muskie. Senator Muskie was the greatest public figure in Maine's history and one of the great legislators in our nation's history. He was the principal author of the 1972 Clean Water Act, which is a cornerstone of our nation's environmental law. He appeared before this committee in 1992 in celebration of the Clean Water Act on its 20th anniversary, and I am honored again to follow in his footsteps.

I focus my remarks today on our progress on the issues that were addressed in the 1987 amendments to the Clean Water Act. As chairman of the Subcommittee on Environmental Protection in that year, I was privileged to manage the bill on the floor of the Senate. That legislation was a heartening example of bipartisan cooperation.

"This committee put it together over a four-year period. Senator Stafford and Senator Quentin Burdick of North Dakota led the committee during that time. I had the pleasure of working on the bill throughout those four years with Senator Dave Durenberger of Minnesota and with Senator John Chafee of Rhode Island. Senator Chafee in 1986 was chairman of the subcommittee and I served as ranking member.

"It is clear beyond doubt that without bipartisan cooperation, the bill would never have become law. I want to join others in especially recognizing Senator Chafee's role as a principal author of what became the Water Quality Act of 1987. I congratulated him on that day 15 years ago and I would like to repeat those words today. Senator Chafee is the architect of this legislation. He chaired the hearings, he managed the bill on the Senate floor, he spoke for the Senate conferees during the long and intense conference with the House. The high quality of this legislation is largely due to his efforts. It is, of course, gratifying that Senator Lincoln Chafee is here today as a member of this committee to continue his father's legacy.

"As I prepared my testimony for this hearing, I was struck by the similarity in the debate over clean water in 1972, 1987 and today. In those early years, we debated the appropriate roles of the Federal Government and the State Governments. We faced opposition to pollution control requirements and implementation schedules. We struggled to find the appropriate level of Federal financial commitment, and we worked to ensure that the Clean Water Act remained relevant to current pollution issues. Each of those concerns remains a vibrant part of today's debate.

"The 1987 amendments can fairly be described as gap-filling measures. We looked at the 1972 law, identified areas where additional action was needed, and sought to create the legal infrastructure needed to further the clean-up of our Nation's waterways. Two key issues in 1987 included funding level and addressing non-point source pollution. There were, of course, many other actions taken in that legislation, such as the creation of the National Estuary Program, the Chesapeake Bay Program, the Great Lakes Program. We reinvigorated the Toxics Program by among other things requiring numerical standards for priority pollutants. We increased the penalties for violations under the Clean Water Act, and we established the first permit program for control of storm water discharges.

A boat ramp along the Merrimack in New Hampshire.

"In 1972, Congress chose to significantly increase Federal participation in clean water programs. It peaked at $5 billion in 1979 and 1980. In 1981, President Reagan proposed the elimination of all funding for clean water unless Congress reduced the size and scope of the program. Congress attempted to respond to the President's demand. Clean water funding was reduced from $5 billion a year to $2.4 billion a year. We reduced the types and numbers of projects that were eligible for Federal funding, and we reduced the Federal share of the cost for construction projects from 75 percent to 55 percent.

"A further step to reform Federal involvement was the adoption of a transition strategy to move the country away from construction grants toward what was then seen as an innovative mechanism called the State Revolving Fund. The 1987 amendments authorized almost $10 billion over 5 years for the phase-out of the construction grants program and $8.4 billion over 5 years for the SRF. We knew at that time that this level of funding was inadequate to fully meet our nation's clean water needs, which then were estimated at between $75 billion and $100 billion. This was a compromise struck between those who favored and those who opposed any Federal investment in clean water. Regrettably, despite our efforts, President Reagan vetoed the bill in 1986. In 1987, Congress reenacted the bill. The President vetoed it again, but this time Congress overrode the veto and the Water Quality Act became law.

"In 1987, we envisioned a situation where after the initial five-year period of Federal investment, the SRF would begin to revolve on its own and the Federal investment in clean water programs would no longer be necessary. That was not the first choice of many of us, but it was necessary to get some legislation enacted to keep the process moving.

Mr. Chairman, as you and the members of the committee know, Federal funding has continued, now at an annual rate of about $1.3 billion a year. I understand that the debate continues over the level of and the mechanism and the formula for distribution of the Federal investment in clean water. There is much debate on that, but there is little or no debate on the need. Just last week, Administrator Whitman announced the results of the EPA's gap analysis, which indicates a gap of over $270 billion for our clean water needs.

"The role of Federal funding in protecting our Nation's waters was at the center of the debate in 1987. It remains there today. In 1987, we knew that we could not possibly fund all that was needed to clean our waters. That is still true. We provided all that we could in 1987 under the circumstances which then existed. You must do so again, because unfortunately, despite all of our efforts, the estimated gap is larger today than it was then. The infrastructure is that much older. Much of it is nearing the end of its useful life, and failure to replace it could threaten public health and our economy.

"I believe the conclusion is clear. Although to act on it will, as always be difficult, there must be an increase in funding for clean water if our nation is to continue its progress in implementing the goals of the Clean Water Act.

"In 1972 and in 1987, the bills survived Presidential vetoes. In each case, cost was a significant issue. In each case, the nation's desire for clean water overshadowed all other issues. I believe that is still the case. The words that Senator Muskie used in 1972 in urging passage of the original Clean Water Act apply to today's challenges, and I would like to quote them for you briefly.

"Senator Muskie said, 'Can we afford clean water? Can we afford rivers and lakes and streams and oceans which continue to make life possible on this planet? Can we afford life itself?' The answers are the same. Those questions were never asked as we destroyed the waters of our nation, and they deserve no answers as we finally move to restore and renew them. The questions answer themselves. We have reached a point in our struggle against water pollution, as we say in New England, "we must either fish or cut bait." If we are serious about restoring the quality of our nation's waters to a level that will support life in the future, then we ought to be prepared to make some sacrifices in that effort now.

Mr. Chairman and members of the committee, I conclude by saying that in 1972 and in 1987 the nation and the Congress rose to meet the challenge. I hope they will do so again.

Thank you, Mr. Chairman, and I will be pleased to answer any questions you may have."

Senator Jeffords. "Thank you so much, Senator Mitchell. It is wonderful to have you here. The message you have given us is one of challenge and one which I certainly believe we should heed and should match your requests.

"I want to thank you for coming, and I want to state that working with you all these years, and having the chance on clean air when you were a real hero on that score, when we needed an upgrading of our air situation, it was one of the most wonderful moments of my life—it wasn't moments; it was weeks, I guess.

Senator Mitchell: "The outcome was wonderful. The process was not, Senator. But I believe even more so, Mr. Chairman. There has been since 1972 a dramatic change in the attitude of the American people toward protection of the environment, fueled by a growing awareness of the threat to the environment that had accumulated over many years prior to that time. I find that when the issues are explained clearly to the American people, the choice of the vast majority is to strongly support protection of our environment and the clean-up of our air and our water. I emphasize again, I do not think this is a partisan issue.

I think the majorities hold largely true across all political, geographic, social and other categories. I believe the American people by overwhelming majorities strongly support the need for the protection of our environment. I can tell you in my own experience in my own State, as Senator Stafford mentioned in Vermont, that the changes that have occurred in the past 30 years have been dramatic, positive, and the people do not want to go back to the days before the Federal Clean Water Act."

Sen. Mitchell continues, "The process that was established initially by Senator Muskie and followed in the 1987 amendments contemplated always a Federal investment and a substantial State role in implementation, with a fairly high level of flexibility at the State level to deal with whatever problems of implementation occurred. I think that one of the factors that has created difficulty is, as you have suggested, the establishment of national standards that go beyond the basic necessities and attempt to resolve every issue in advance, which I think cannot be done in a country of the size, diversity and competing interests as large as this one.

"Now, that is an easy formulation to state and very difficult to implement because in the minds of each of us here, what is or is not essential as a Federal standard may differ, and how much flexibility for the State and local governments will also differ in the minds of each person. It is in that area that I believe the greatest contribution can come from members of this committee. When we did this in 1987, as both Senator Stafford and I have noted, it took four years. John Chafee, Dave Durenberger, Bob Stafford, myself, Quentin Burdick and a few other members of the committee worked at it over a very long period of time through debate, discussion, trial, error—trying to find the right process of formulation. "

Sen. Mitchell continues, "We encountered a difficulty that I hope you do not encounter, and that was, of course, the President's demand for a complete end of the program. We struggled to keep the program alive in a way that we hoped would meet the President's approval, even though we believed it ought to be much more than it was at the time. I do not think anyone can—I know I cannot, and I am not sure anyone can be more precise than that in response to the question—but on the Safe Drinking Water Act,

"Senator, I come from Maine where many small towns had precisely that problem. What we found was—in dealing with it in Maine—is that there had to be a substantial degree of flexibility in dealing with particular problems because so much of this depends upon local circumstance.

It is, as I repeat now, the greatest contribution this committee can make is in finding the appropriate balance between a broad Federal mandate supported by substantial Federal investment, and a sufficiently high degree of flexibility at the State and local level so you do not get decisions that appear to ordinary American citizens as contrary to common sense, which is what happened in the case of some of the application of the Safe Drinking Water Act and in other areas of environmental regulation. I think it all has to pass in the minds of the average American a commonsense test. I think you will agree, and I think it is clear beyond dispute, that there is a broad reservoir of support in this country for meaningful, sensible environmental legislation to protect and to enhance our Nation's waters and air."

The Merrimack is a dominant resource near Haverhill and other parts of the North Shore of Massachusetts. (*DigitalMassachusetts photo*)

20

Timeline of a Greener America

In the year 2020, most American adults are familiar with the environmental movement. At the least, we expect the federal government to play a leadership role in maintaining a clean Merrimack, a glistening Atlantic, and wetlands that support wildlife of all kinds.

Yet this was not always the case. I was surprised to learn that serious legislation did not start until the early 1970s after the Cuyahoga River in Ohio caught fire. *Time* used a photo on a cover in 1969 showing the river with shocking flames and billowing smoke. That photo horrified Americans and stimulated people to work for cleaner rivers, but the photo itself was taken at the site more than a decade earlier when a railroad depot adjacent to the river was ablaze. This alarming conflagration that "ignited" indignation about water pollution did not actually represent the situation—wrong photo, right response.

The following is a timeline of key events in recent decades that have resulted in a framework through which to work for clean water and air. It was compiled by the Environmental Protection Agency.

September 27, 1962: Rachel Carson's *Silent Spring* Published

Rachel Carson's *Silent Spring*, a critical look at pollution in the United States, jumpstarts the environmental movement. Carson, a birdwatcher, discovered that heavy use of pesticides was killing off birds and making the forests "silent." She also wrote about the harmful chemicals used for defoliation in the Vietnam War.

June 29, 1969: Fire on the Cuyahoga River

In 1969, the Cuyahoga River in Ohio becomes so polluted that it catches on fire. The fire helped spur an avalanche of water pollution control activities such as the Clean Water Act and the Great Lakes Water Quality Agreement. By bringing national attention to water pollution issues, the Cuyahoga River fire was one of the events that led to the creation of the federal Environmental Protection Agency and the Ohio Environmental Protection Agency.

Sunset over the Merrimack often provides scenic moments. (*Dan Graovac photo*)

The falls in Manchester provides power for thousands.

1970: President Nixon Signs NEPA

The bill forms the Council on Environmental Quality (CEQ) to advise the president on the environment and review federal agencies' environmental impact statements, required for projects that would affect the environment.

April 22, 1970: First Earth Day

More than 20 million Americans participate in one of the largest grassroots community service movements in our history. Earth Day is now celebrated every year by almost 1 billion people worldwide.

December 2, 1970: Official Formation of EPA

After President Richard Nixon's "Reorganization Plan No. 3" issued in July 1970, EPA is established on December 2, 1970. The agency consolidates federal research, monitoring, and enforcement activities in a single agency. EPA's mission is to protect human health by safeguarding the air we breathe, the water we drink, and the land on which we live.

December 31, 1970: Clean Air Act of 1970

Congress authorizes EPA to set national air quality, auto emission, and anti-pollution standards. The standards led to the production of the catalytic converter in 1973 by New Jersey's Engelhard Corporation. In its first twenty years, the Clean Air Act prevented more than 200,000 premature deaths by significantly reducing the presence of lead, sulfur dioxide, and other harmful pollutants in the air.

December 31, 1971: Vehicle Fuel Economy Testing

EPA begins testing the fuel economy of cars, trucks, and other vehicles—the first step toward informing consumers about the gas mileage of their vehicles.

June 14, 1972: EPA Bans DDT

After public concerns about the health effects of the widely used pesticide DDT, EPA bans its use and requires an extensive review of all pesticides.

October 18, 1972: Clean Water Act

Congress passes the Federal Water Pollution Control Act, commonly known as the Clean Water Act. The purpose of the Clean Water Act is to restore and maintain our nation's waters by preventing pollution, providing assistance to publicly owned wastewater treatment facilities, and maintaining the integrity of wetlands.

October 23, 1972: Ocean Dumping Act

Congress enacts the Marine Protection, Research, and Sanctuaries Act, or Ocean Dumping Act, to reduce ocean water pollution. Within three years, EPA had denied seventy contracts, many of them for chemical dumping.

June 4, 1973: Auto Maintenance Regulations

EPA sets regulations on car manufacturing and testing for compliance with Clean Air Act emissions standards. Car manufacturers were required to install malfunction warning systems in cars made in 1975 or later, and EPA specified conditions under

Transport vessels are lined up at Cashman Park, Newburyport.

which manufacturers could perform maintenance to cars during their 50,000-mile test for meeting emissions standards.

July 1973: Oil Embargo
The OPEC oil embargo triggers a spike in oil prices and scarcity of supply, stimulating conservation and research into alternative energy sources.

December 28, 1973: Leaded Gasoline Phase-Out
EPA released a study confirming that lead from automobile exhaust posed a direct threat to public health. On December 8 of that same year, EPA issues regulations gradually reducing lead in gasoline.

December 16, 1974: Safe Drinking Water Act
Congress passes the Safe Drinking Water Act, allowing EPA to regulate the quality of public drinking water.

June 25, 1977: Safer Drinking Water
National drinking water standards went into effect for the first time. For the first time, all public water suppliers were required to test their public water routinely and notify their customers if water was not up to EPA standards.

JUNE 25, 1977: SAFER DRINKING WATER

National drinking water standards went into effect. For the first time, all public water suppliers were required to test their public water routinely and notify their customers if water was not up to EPA standards.

AUGUST 8, 1977: CLEAN AIR ACT AMENDMENTS

President Jimmy Carter signs the Clean Air Act Amendments to strengthen air quality standards and protect human health. The stronger protections spurred the development of scrubber technology, which removes air pollution from coal-fired power plants.

DECEMBER 28, 1977: CLEAN WATER ACT OF 1977

President Jimmy Carter signs the Clean Water Act, amending the 1972 version. The act stressed the importance of toxic pollutant control. A construction grant over five years created thousands of jobs, aided state and local planning, and encouraged experimentation with new water treatment methods.

AUGUST 2, 1978: LOVE CANAL DISASTER

Residents discover that Love Canal, NY, is contaminated by leaking chemical containers. The pollution is linked to serious health threats such as cancer and birth defects. President Carter declares an emergency, authorizing EPA to help temporarily relocate about 700 families. In 1980, Congress passed Comprehensive Environmental Response, Compensation, and Liability Act, also called the Superfund Act, authorizing EPA to identify those responsible for the contamination and compel them to clean up the sites.

DECEMBER 10, 1980: CONGRESS CREATES THE SUPERFUND PROGRAM

Congress creates the Superfund Program, holding polluters responsible for cleaning up most hazardous waste sites. The initial program designated $1 billion for cleanup efforts.

SEPTEMBER 15, 1982: ENVIRONMENTAL JUSTICE MOVEMENT BEGINS

A PCB landfill protest in Warren County, North Carolina—a predominantly poor, African American area—launches the environmental justice movement. Environmental justice is the fair treatment and involvement of all people, regardless of race or income, in decisions on development, implementation, and enforcement of environmental policies.

MARCH 4, 1985: 90 PERCENT REDUCTION OF LEAD IN GASOLINE

EPA sets the final standard to 0.1 g per leaded gallon starting January 1, 1986. An interim standard of 0.5 g per leaded gallon took effect on July 1, 1985.

OCTOBER 6, 1986: WETLANDS PROTECTION

Administrator Thomas names endangered wetlands as a top EPA priority and announces the new Office of Wetlands Protection. He charged the new office with researching wetlands ecosystems, reaching out to property owners and educating them on wetlands value, and strengthening protection measures.

NOVEMBER 21, 1988: SEWAGE OCEAN-DUMPING BAN

Congress bans ocean-dumping of sewage sludge and industrial waste.

Voyagers paddle through Haverhill on their four-day trip in 2019. (*Bryan Eaton photo*)

April 19, 1990: Toxic Release Inventory (TRI)

EPA launches the toxic release inventory, which tells the public which pollutants are being released from specific facilities in their communities.

November 15, 1990: Acid Rain Controls Enacted

Congress passes the Clean Air Act (CAA) Amendments, imposing a new cap and trade approach to addressing acid rain by reducing sulfur dioxide emissions from power plants across the country. The amendments also implement controls to phase out ozone-depleting substances, like CFCs. In their place, industry creates new technology that is cleaner, cuts costs, and improves productivity and quality.

June 30, 1992: New York City Stops Dumping

New York City was the last U.S. city to dump sewage at sea. It stopped as a result of its agreement with EPA under the Ocean Dumping Ban Act of 1988.

June 30, 1994: Brownfields Program

EPA launches its Brownfields Program to clean up abandoned, contaminated sites and return them to productive community use. In the years since brownfields began, EPA has cleaned up more than 450 contaminated sites. Through Brownfields job-training partnerships, the program has played a role in creating more than 61,000 jobs.

January 29, 1996: Leaded Gasoline Phased Out

EPA completes a twenty-five-year mission to completely remove lead from gasoline. Administrator Browner called it one of the great environmental achievements of all time.

DECEMBER 19, 1995: MUNICIPAL WASTE COMBUSTORS (MWCs)

EPA requires municipal waste combustors—incinerators that burn household and commercial waste—to reduce toxic emissions by 90 percent from 1990 levels. About 90 percent of municipal waste combustors are waste-to-energy plants that generate electricity or steam from burning garbage for commercial and residential use.

AUGUST 6, 1996: SAFE DRINKING WATER ACT (SDWA) AMENDMENTS

President Clinton signs amendments to the Safe Drinking Water Act to improve protections ensuring that all Americans have access to clean, safe drinking water. The law makes funding available to upgrade water treatment plants and requires public drinking water suppliers to inform customers about chemicals and microbes in their water.

FEBRUARY 2002: HUDSON RIVER CLEANUP

EPA officially moves to clean up Hudson River PCB contamination by removing approximately 2.65 million cubic yards of contaminated sediment from a 40-mile stretch of the river.

JUNE 12, 2006: EPA'S WATERSENSE PROGRAM CREATED

EPA and industry partners join to create the WaterSense Program, which seeks to protect the future of our nation's water supply by offering practical ways to use less water. Through the use of water-efficient products and services, WaterSense has helped consumers save 46 billion gallons of water and $343 million on their water and sewer bills since its launch.

DECEMBER 7, 2009: GREENHOUSE GASES ENDANGERMENT FINDING

After a thorough examination of the science and careful consideration of public comments, EPA announces that greenhouse gases threaten the health and welfare of the American people. As a result, greenhouse gases that lead to climate change can be regulated under the Clean Air Act.

MAY 27, 2015: CLEAN WATER RULE PROTECTS CRITICAL STREAMS AND WETLANDS

The rule, later retitled the Waters of the U.S. (WOTUS) Rule, protects critical streams and wetlands by ensuring that waters protected under the Clean Water Act are more precisely defined and predictably determined.

JUNE 2017: U.S. WITHDRAWS FROM THE PARIS CLIMATE ACCORD

President Trump orders the cessation of all implementation of the accord, stating that compliance with the terms of the agreement could undermine U.S. competitiveness and cost the U.S. as much as 2.7 million lost jobs by 2025.

JUNE 27, 2017: EPA, ARMY MOVE TO RESCIND 2015 "WATERS OF THE U.S." RULE

To provide regulatory certainty, EPA proposes rules to rescind the Waters of the U.S. Rule (formerly the Clean Water Rule) and redefine the term "waters of the United States."

OCTOBER 10, 2017: EPA PROPOSES REPEAL OF THE CLEAN POWER PLAN

EPA proposes to repeal the Clean Power Plan after determining that the Obama-era regulation exceeds the Agency's statutory authority.

21

Modern Voices
on the River

Lori Trahan: Working for a Cleaner River

Congresswoman Lori Trahan represents the Lowell area and has been an advocate for federal funding to improve sewage-treatment plans along the Merrimack. In past years, low-interest federal loans have been available to communities. EPA officials say they are a bargain because the interest rate does not rise with inflation. It will always stay low even while costs increase, but officials in the Trump administration have not been supportive of spending federal dollars for clean rivers, lakes, and oceans.

Trahan speaks frequently about the need for clean rivers. Before the pandemic, she addressed a group in Chelmsford that had been assembled by the Merrimack River Watershed Council.

"In Washington, they call me the sewage lady," Trahan said about her clean-water activities. "There are so many issues in Congress at one time, and because I have been working hard on this issue, I am associated with this cause." She said the river was a key asset to dozens of communities in Massachusetts and New Hampshire.

The Merrimack River began improving following the Clean Water Act of 1972. It provided grants to scores of communities, and this money enabled them to build sewage-treatment plants. It is unnerving to think that many communities were dumping raw sewage into the river even into the 1970s, and old-timers who grew up near the river can remember its odor.

Older residents of the North Shore remember when the river was a threat, not a resource. Dick Canepa, a Newburyport octogenarian who taught for decades at Pentucket High School, said his parents did not want him to go near the river in the 1950s and '60s.

The Clean Water Act helped the Merrimack River, as well as many others. The keyword is "grant." Communities would get a federal "gift" that covered most costs of a treatment plant, but in 1987, the Reagan Administration promulgated legislation that made the

Congresswoman Lori Trahan, a Democrat from the Lowell area, has worked to acquire low-interest loans for communities so they can improve their sewage-treatment systems.

program go from "grants" to "loans." Grants were essentially free; loans must be repaid and are less attractive to municipal leaders.

Trahan is seeking loans from the federal Clean Water State Revolving Fund. That is still classified as a low-interest loan, but it would enable communities to begin work on their plants. One of the first steps would be to develop a program that would separate storm water from effluent. That would be very expensive. Also, many sewer lines are half a century old and need to be replaced.

Yet Trahan has made a clean river a priority. Trahan is the first in her family to graduate from college. She earned a scholarship to play Division 1 volleyball at Georgetown University. Thereafter, she joined the staff of former Congressman Marty Meehan as a scheduler, eventually working her way up to chief of staff.

Following her public service, Trahan began working in the private sector as the only female executive at a tech company before moving on to cofound a woman-owned and operated consulting firm, where she advised various companies on business strategy. She taught companies how to create the conditions for employees—especially women—to thrive.

As a member of the House Education and Labor and House Armed Services Committees, Trahan said she is focused on fighting for working families on issues such as affordable health care, quality public education, workforce development, the environment, and working to end the opioid crisis.

There is also always the pesky problem of improving the quality of the Merrimack River.

STATE SEN. DIANA DiZOGLIO: A LEADER IN CREATING THE MERRIMACK RIVER DISTRICT COMMISSION

This state senator has been among the most active in pressing for a cleaner river. DiZoglio, a Democrat from Methuen, was a prime mover of an initiative to create the Merrimack River District Commission in 2020. This is an organization made up of representatives from numerous groups. It is dedicated to cleaning and improving the river. With a grant from the state Legislature, she has led the battle to preserve the Merrimack.

DiZoglio showed her commitment in the summer of 2019. She was part of the Voyagers team that kayaked the river from Franklin, N.H., to Newburyport. She was one of perhaps a dozen who paddled on every leg of the journey. That included getting drenched by evening rain and by having to seek shelter in distant communities.

It brought enormous media attention to the Merrimack. The Voyagers met with the media at the end of each day's paddling. Reporters from radio, TV, newspapers, and online services from almost every community along the waterway provided coverage. This was a useful activity because some North Shore residents do not seem to have a clear picture of what is wrong with the river. News stories appear and then are gone. The Voyagers were able to explain the causes of pollution and warn that steps must be taken to limit the number of CSOs (combined sewage overflows). The Voyagers traveled before the pandemic, of course, and many residents said that a clean river is an important goal for those on the North Shore.

DiZoglio won the first Essex District seat in 2018, and the district includes Newburyport, Methuen, Haverhill, Merrimack, Amesbury, Salisbury, and parts of North Andover. When she was a state representative from the fourteenth Essex District, her zone included North Andover, Methuen, Lawrence, and Haverhill.

She indicated that representing the communities that rely on summer tourism, including Newburyport, Salisbury, and Amesbury, has made her more aware of the economic value of the river to these communities.

DiZoglio was born in Methuen and graduated from Methuen High School in 2002. She attended Wellesley College, graduating with a degree in psychology and Spanish. Prior to being elected to the Massachusetts House of Representatives, DiZoglio worked as chief-of-staff to Edward A. Kelly, president of the Professional Fire Fighters of Massachusetts. She also served as a legislative aide in the Massachusetts House of Representatives, worked for multiple non-profit organizations and was a small business owner. DiZoglio was elected to the Massachusetts House of Representatives in 2012, running a successful primary challenge to incumbent state representative David M. Torisi.

DiZoglio said the river has enormous meaning in the Merrimack Valley:

> It's a great natural resource and has so many beautiful parts … It appeals to so many people. There is a boathouse in Lawrence that offers kayaking, sailing and other things for people in that city. People can learn; they can get out on the river. That's just one example of the value of the Merrimack.

DiZoglio, like others who have worked on legislation focusing on the river, said that the federal delegation from New England is trying to obtain funds through the federal Clean Water State Revolving Fund. In its own words, this is a "federal-state partnership that

State Sen. Diana DiZoglio, D-Methuen, has been a leader in uniting local officials to take better care of the river. (*Dan Graovac photo*)

The Voyagers trip down the Merrimack brought surprises, including meeting a herd of cattle. In the bow of this kayak is state Sen. Diana DiZoglio. (*Dan Graovac photo*)

provides communities a permanent, independent source of low-cost financing for a wide range of water-quality infrastructure projects." DiZoglio said in an interview in 2020:

> Our trip on the river revealed so many beautiful parts of the Merrimack. Since I've become a state senator after being a state representative, the people in my senate district are really committed to a clean river. It's going to take time but the commitment to preserve the river is strong.

JAMES KELCOURSE: STATE REPRESENTATIVE WHO GREW UP ON THE MERRIMACK

State Rep. James Kelcourse is another elected official who was on the Voyagers trip. Kelcourse has a remarkably strong tie to the Merrimack. His father, Larry Kelcourse, owned a riverside marina in Amesbury for several decades before selling it in 2019.

Jim is a resident of Amesbury, and he represents Amesbury, Newburyport, and Salisbury. The Republican is a former Amesbury city councilor. He defeated Newburyport City Councilor Ed Cameron in 2016 by twelve votes after a month-long recount. Kelcourse, a graduate of Villanova University, said:

> The river means so much to me … I worked in the marina when I was a kid, helping people with their boats and helping my father with his business. I spent a lot of time fishing in the river and then going out to sea, for stripers, cod, flounder, everything. It was great.
>
> The river was dirty when I grew up. My father, most parents really, did not let their kids swim in the Merrimack. In the old days, when there were factories on the river, the mills would empty their trash into the Merrimack, and they say the river would turn the color of the dyes that were released.
>
> It's so much better now. The river means everything to this area. There are a lot of fishing boats and now there so many restaurants, where people can enjoy themselves. We have an eagle festival, we have fishing, boating and sailing; the river offers so much. The Clean Water Act (1972) was the start of the improvement.
>
> Many of us are working hard to keep it clean and safe. It's going to take a lot of money but there is real commitment in our area. I have fine memories from when I was a kid—it's great to see it cleaner and I am going to work to keep it that way.

Mayor Donna Holaday of Newburyport has been a key supporter of the river. Under her leadership, the city improved its treatment plant along the Merrimack. City officials even added expensive anti-odor hardware at the request of residents who lived near the plant.

She has expressed concern that communities "downriver" are the recipients of CSOs when they flow east. In addition to advocating for the river itself, Holaday engineered an agreement that helped Plum Island, which is adjacent to the Merrimack. When builders installed a faulty system to take sewage from the island to the city's treatment plant, she obtained a settlement of close to $5 million so the system could be fixed.

Another elected official who has worked to help the Merrimack River is state Sen. Bruce Tarr, R-Gloucester. With Salisbury public official Jerry Klima, he has been a leader of the Merrimack River Beach Alliance. This organization has met regularly to develop a strategy to keep the river navigable for boats. Another goal was to limit erosion of the beach, which was making some houses vulnerable to falling into the ocean.

Newburyport Mayor Donna Holaday has been
responsible for major upgrades of the city's
sewage treatment plant, and for urging cities
upstream to cut down on their effluent emissions.

Between 2013 and 2016, the alliance helped shake free close to $20 million in federal funds to fortify the two jetties at the point where the Merrimack meets the Atlantic. Tarr runs a very effective meeting. The alliance would draw scores of officials, state employees, municipal department heads, Coast Guard leaders, and waterfront residents. Members of the EPA and the state environmental organizations also cooperated in plans to improve the area. This bi-partisan organization achieved a great deal in several years. The Jetties project is a major achievement, and leaders hope that it makes the river more navigable.

Heather McCann is executive director of Groundwork Lawrence. She traveled with the Voyagers and is hopeful that the river can remain clean enough for recreation:

I grew up on the Merrimack in New Hampshire, and we couldn't really use the river …
It was too dirty. In recent years it has become clean enough for kayaking, fishing and
boating. We must do everything we can to keep it the great resource that it is.

Dianne Sherratt-Steimel is a journalist who produced a quality film about the river titled *Troubled Waters*. Her headquarters are in Chelmsford with a cable-news operation, and her film shows the beauty—and the challenges—of the Merrimack:

It's such a great resource … There are great recreational opportunities, and in most
communities, you can walk along the Merrimack and enjoy the beauty.

Most legislators are aware of its vulnerability. One thing that's surprised me is how much
money and effort it is going to take to stop the combined sewage overflows along the river.

Numerous cities discharge waste after heavy rains. And the EPA allows it to happen.
If it didn't go into the river, it would back up in the sewer system, and get into homes
and businesses. No one wants that.

It will be expensive. And we're not just talking about the Merrimack. There are many rivers
in this country that have the same problem. But along the Merrimack, people are concerned.

It will be very expensive. But more people along the river are becoming aware of it,
and I think people will consider supporting a clean river.

22

Lane Glenn, President of Northern Essex Community College, Led the Voyagers

Lane Glenn was a key organizer of the Voyagers and is a robust adventurer. The remainder of this chapter is his writing about his journey from Franklin to Newburyport in 2019.

The ebbing and flowing of the Merrimack River has been the heartbeat of New England life for centuries. From colonial fishing and farming villages, to the mills built on its banks in the earliest days of America's Industrial Revolution, to the recreational boathouses and marinas that line the shore today, the Merrimack draws us closer to the power, the beauty, and the abundance of nature.

The Merrimack also offers the nearly 700,000 people who live in the 65 cities and towns along its banks in Massachusetts and New Hampshire a common thread and shared purpose—a way to feel and to be part of a community. While there are many things that may separate us—politics, wealth, education, and more—the Merrimack brings us together.

It was this spirit of common purpose and togetherness that formed the Merrimack River Valley Voyagers, a group of local leaders who in 2019 spent four days paddling and camping along the 117-mile length of the Mighty Merrimack, enjoying its beauty and bringing attention to its environmental needs, and economic, educational, and recreational opportunities.

I had the privilege of helping to organize the core group of Voyagers, who kayaked the full length of the river, including state Sen. Diana DiZoglio, state representatives Jim Kelcourse and Christina Minicucci; Dougan Sherwood, president of the Greater Haverhill Chamber of Commerce; Derek Mitchell, executive director of the Lawrence Partnership; Dan Graovac, president of the Merrimack River Watershed Council Board of Directors;, and Heather McMann, executive director of Groundwork Lawrence.

In addition to this adventurous lot, a number of other community leaders in both New Hampshire and Massachusetts jumped in boats and joined us for a day or two along the river, including state Sen. Edward Kennedy; state representatives Linda Dean Campbell and

Andy Vargas; Newburyport Mayor Donna Holaday, and Newburyport City Councilors Barry Connell and Charles Tontar.

We managed to draw plenty of attention to our trek, with ample social media coverage, prominent stories in newspapers in the major cities along the river, interviews and updates on NPR, spots on local cable television, and a feature in the fall issue of *Merrimack Valley Magazine*, thanks to editor Doug Sparks, who paddled along and camped with us for part of our journey.

They were four incredible days. Each of us has tales to tell and I encourage you to ask the Voyagers to share some of their favorites when you see them.

Here are just a few of mine:

We launched our kayak flotilla in Franklin, New Hampshire, where the confluence of the Pemigewasset and Winnipesaukee rivers form the headwaters of the Merrimack, early in the morning on Aug. 7, 2019. A few of us were experienced kayakers, and some were new to paddling. The northernmost part of the river is mostly wide with a gentle current, giving everyone a chance to get their sea legs underneath them for the first few hours.

The first 20 or so miles are also some of the most rural and least developed portions of the river, and we enjoyed the company of geese, egrets, great blue herons, and a pair of eagles that soared overhead, fishing for their breakfast.

The birds we all expected, but the bovines we did not. Somewhere south of Penacook, N.H., we were greeted by a herd of cows that came running (really, not making this up) through a grove of trees on the riverbank, down into the water for a midday dip.

We had planned a late afternoon press conference with Congresswoman Annie Kuster at the New Hampshire Technical Institute south of Concord, but thunderstorms drove us off the river early. Instead of camping where we had planned, at the White Sands Conservation Area in Pembroke, we found shelter from the storm at the Push-Me-Pull-You Llama Farm in nearby Canterbury.

Refreshed and ready to roll down the river the next morning, we met Thomas O'Donovan, director of the New Hampshire Department of Environmental Services Water Division at our launch site to learn more about one of the big challenges facing the Merrimack River today.

Although the river is much cleaner than it was a generation or more ago, when industrial waste and household sewage had turned it into a toxic flowing cesspool, in 2016 it still made American Rivers' annual list "America's Most Endangered Waterways." This was mostly due to the large amounts of combined sewage overflow, or CSOs, that are dumped into the river after large storms—like the one we experienced on our first afternoon.

The Merrimack River Watershed Council estimates that, last year, at least 770 million gallons of untreated sewage was discharged into the Merrimack from urban sewage treatment systems in Manchester, Nashua, Lowell, Lawrence, and Haverhill. While all of these cities are taking steps to limit and eventually end CSO discharge, costs run into the hundreds of millions of dollars, and it will take decades or longer to accomplish.

Just two days before we set out on our journey, the Massachusetts State Senate passed a bill based on legislation originally filed by one of our Voyagers, Sen. Diana DiZoglio, D-Methuen. This bill would create a Merrimack River District Commission comprised of state and local officials, and members of regional and environmental groups, to comprehensively study the river and create a regional plan for addressing environmental concerns.

Energetic Voyagers kayaked the 117-mile Merrimack in 2019. From left are state Sen. Diana DiZoglio; Derek Mitchell, executive director, Lawrence Partnership; Lane Glenn, president of Northern Essex Community College; Heather McMann, executive director, Groundwork Lawrence; state Rep. Jim Kelcourse; state Rep. Christina Minicucci; Dan Graovac, board chair of the Merrimack River Watershed Council; Dougan Sherwood, president, Greater Haverhill Chamber of Commerce; and Doug Sparks, editor of Merrimack Valley Magazine. (*Northern Essex Community College photo*)

Lane Glenn, a leader of the Voyagers, concentrates on a run through some rapid water. (*Dan Graovac photo*)

Just a few miles downstream on Day 2, we pulled our boats out of the water to portage around the Hooksett Falls Dam, an operation requiring nine paddlers to haul eight boats (seven singles and a tandem) a quarter mile or so along the riverbank, over some railroad tracks, and launch them back in the stream again on the other side.

It was also the first opportunity for some of the daredevils in our paddling pack to try to run the rapids and smaller falls at the foot of the Hooksett Dam—and for others to wait downstream and catch their gear before it floated away after a tumble or two!

Packing, paddling, and portaging together really brought out the best in everyone, including this marvelous show of bipartisan crewing as Republican Representative Jim Kelcourse took a turn in the tandem with Democratic Senator Diana DiZoglio—working together to get things done for Massachusetts!

After a public event and press conference at the Amoskeag Fish Ladder Dam in Manchester, N.H., highlighting some of the environmental concerns for the Merrimack, the Voyagers had a chance to portage again. Then we ran the five-mile urban canyon stretch of rock-strewn rapids through Manchester on route to the Independence Rowing Club in Nashua, N.H., our camp site for the evening (after another thunderstorm chased us off the river a few miles short of our targeted landing spot).

Day 3 was a big one for us, as we paddled our way into Lowell for a press conference at the UMass Lowell Bellegarde Boathouse, followed by a trailer portage around Pawtucket Falls, another launching behind the Lowell Spinners Stadium, and an evening landing with a friendly hometown crowd at Greater Lawrence Community Boating.

Heather McMann of Groundwork Lawrence and Derek Mitchell of the Lawrence Partnership led our press conference at the Abe Bashara Boat House. We talked about the magnificent contributions GWL has made to cleaning up the river, as well as economic and workforce development opportunities along the Merrimack.

And I shared this proud fact with the crowd: Northern Essex Community College has nearly 40,000 active alumni, and more than 25,000 of them live along the Merrimack River today—in every single city and town in the Merrimack Valley, from Franklin south through Concord, Manchester, Nashua, Lowell, Lawrence, Haverhill, and out to Newburyport, Salisbury, and Plum Island.

Just below the Stone Dam in Lawrence, the Merrimack River rises and falls as much as nine feet, with the tide pushing in through the mouth between Salisbury and Plum Island and flowing back out again into the Atlantic Ocean.

Our destination on Day 4 was 27 miles downriver, where low tide at Plum Island was at 2.30 in the afternoon. We wanted to ride the tide out to the beach, and definitely not get caught paddling against it across Newburyport Harbor.

That meant a 6 a.m. launch from Pemberton Park in downtown Lawrence, as the sunlight was just beginning to spill across the mill buildings that line the riverbanks.

By 9 a.m. we were a couple of miles outside of Haverhill, where Jessica Winiecki, a member of the city's Harbor Commission and River Access Committee, helpfully provided homemade Sunday brunch for the Voyagers dockside behind her house on the river.

After landing and docking our boats in Haverhill, we walked along Merrimack Street downtown, stopping by the construction of The Heights, future home to NECC's Hospitality and Culinary Arts Institute. And we attended a public event led by Greater Haverhill Chamber of Commerce by President Dougan Sherwood behind Harbor Place, the five-story building at the Basiliere Bridge that is helping to transform Haverhill's riverfront.

With no time to lose if we wanted to beat the turning of the tide, we hopped back in our kayaks for our final push to the sea. Just as we paddled across Newburyport harbor, a squall kicked up, blowing heavy winds and some pelting rain as our boats began to land, one at a time, on the sand at Captain's Landing on Plum Island.

The journey of the Merrimack River Valley Voyagers was a physical accomplishment: 117 miles, 30 miles a day, with five portages, several rapids runs, two full-blown thunderstorms, and a sea squall. We paddled, carried kayaks, pitched tents, slept on the ground, baked in the sun, and soaked in the rain.

The Voyagers' journey was also a public awareness accomplishment: We brought a tremendous amount of attention to the environmental challenges the river faces, and the policies, politics, and economics of responding to those challenges.

And most of all, our journey was a community accomplishment, creating and deepening relationships among our hardy band of voyagers, and with friends and fellow travelers all along the way, where this river we now all know so much better endlessly ebbs and flows, the beating heart of the Merrimack Valley.

23

POLLUTION:
THE NATIONAL RESOURCES DEFENSE COUNCIL EXPLAINS THE PROBLEM AND OFFERS A CALL TO ACTION

This essay was produced by the National Resources Defense Council several years ago and provides an overview of pollution and clean rivers. It is included here so readers will have a common ground from which to understand the challenge.

British poet W. H. Auden once noted, "Thousands have lived without love, not one without water." Yet while we all know water is crucial for life, we trash it anyway. Some 80 percent of the world's wastewater is dumped—largely untreated—back into the environment, polluting rivers, lakes, and oceans.

This widespread problem of water pollution is jeopardizing our health. Unsafe water kills more people each year than war and all other forms of violence combined. Meanwhile, our drinkable water sources are finite: Less than 1 percent of the earth's freshwater is actually accessible to us. Without action, the challenges will only increase by 2050, when global demand for freshwater is expected to be one-third greater than it is now.

Sip a glass of cool, clear water as you read this, and you may think water pollution is a problem … somewhere else. But while most Americans have access to safe drinking water, potentially harmful contaminants—from arsenic to copper to lead—have been found in the tap water of every single state in the nation.

Still, we're not hopeless against the threat to clean water. To better understand the problem and what we can do about it, here's an overview of what water pollution is, what causes it, and how we can protect ourselves.

Kayaks belonging to the Voyagers lined the river in summer of 2019, ready to start another day of traveling downstream. (*Dan Graovac photo*)

What Is Water Pollution?

Water pollution occurs when harmful substances—often chemicals or microorganisms—contaminate a stream, river, lake, ocean, aquifer, or other body of water, degrading water quality and rendering it toxic to humans or the environment.

What Are the Causes of Water Pollution?

Water is uniquely vulnerable to pollution. Known as a "universal solvent," water is able to dissolve more substances than any other liquid on earth. It's the reason we have Kool-Aid and brilliant blue waterfalls. It's also why water is so easily polluted. Toxic substances from farms, towns, and factories readily dissolve into and mix with it, causing water pollution.

Categories of Water Pollution

Groundwater

When rain falls and seeps deep into the earth, filling the cracks, crevices, and porous spaces of an aquifer (basically an underground storehouse of water), it becomes groundwater—one of our least visible but most important natural resources. Nearly 40 percent of Americans rely on groundwater, pumped to the earth's surface, for drinking water. For some folks in rural areas, it's their only freshwater source. Groundwater gets polluted when contaminants—from pesticides and fertilizers to waste leached from landfills and septic systems—make their way into an aquifer, rendering it unsafe for human use. Ridding groundwater of contaminants can be difficult to impossible, as well as costly. Once polluted, an aquifer may be unusable for decades, or even thousands of years. Groundwater can also spread contamination far from the original polluting source as it seeps into streams, lakes, and oceans.

Surface Water

Covering about 70 percent of the earth, surface water is what fills our oceans, lakes, rivers, and all those other blue bits on the world map. Surface water from freshwater sources (that is, from sources other than the ocean) accounts for more than 60 percent of the water delivered to American homes. But a significant pool of that water is in peril.

According to the most recent surveys on national water quality from the U.S. Environmental Protection Agency, nearly half of our rivers and streams and more than one-third of our lakes are polluted and unfit for swimming, fishing, and drinking. Nutrient pollution, which includes nitrates and phosphates, is the leading type of contamination in these freshwater sources. While plants and animals need these nutrients to grow, they have become a major pollutant due to farm waste and fertilizer run-off. Municipal and industrial waste discharges contribute their fair share of toxins as well. There's also all the random junk that industry and individuals dump directly into waterways.

Newburyport has several large pipes that send rainwater into the river. This one is located at the west end of the Peter Matthews Memorial Boardwalk.

Ocean Water

Eighty percent of ocean pollution (also called marine pollution) originates on land—whether along the coast or far inland. Contaminants such as chemicals, nutrients, and heavy metals are carried from farms, factories, and cities by streams and rivers into our bays and estuaries; from there they travel out to sea. Meanwhile, marine debris—particularly plastic—is blown in by the wind or washed in via storm drains and sewers. Our seas arc also sometimes spoiled by oil spills and leaks—big and small—and are consistently soaking up carbon pollution from the air. The ocean absorbs as much as a quarter of man-made carbon emissions.

POINT SOURCE

When contamination originates from a single source, it's called point source pollution. Examples include wastewater (also called effluent) discharged legally or illegally by a manufacturer, oil refinery, or wastewater treatment facility, as well as contamination from leaking septic systems, chemical and oil spills, and illegal dumping. The EPA regulates point source pollution by establishing limits on what can be discharged by a facility directly into a body of water. While point source pollution originates from a specific place, it can affect miles of waterways and ocean.

NONPOINT SOURCE

Nonpoint source pollution is contamination derived from diffuse sources. These may include agricultural or stormwater run-off or debris blown into waterways from land. Nonpoint source pollution is the leading cause of water pollution in U.S. waters, but it's difficult to regulate, since there's no single, identifiable culprit.

TRANSBOUNDARY

It goes without saying that water pollution can't be contained by a line on a map. Transboundary pollution is the result of contaminated water from one country spilling into the waters of another. Contamination can result from a disaster—like an oil spill—or the slow, downriver creep of industrial, agricultural, or municipal discharge.

THE MOST COMMON TYPES OF WATER CONTAMINATION

Agricultural

Not only is the agricultural sector the biggest consumer of global freshwater resources, with farming and livestock production using about 70 percent of the earth's surface water supplies, but it's also a serious water polluter. Around the world, agriculture is the leading cause of water degradation. In the United States, agricultural pollution is the top source of contamination in rivers and streams, the second-biggest source in wetlands, and the

A pipe in Methuen is near the site of debris and garbage. (*Dan Graovac photo*)

third main source in lakes. It's also a major contributor of contamination to estuaries and groundwater. Every time it rains, fertilizers, pesticides, and animal waste from farms and livestock operations wash nutrients and pathogens—such as bacteria and viruses—into our waterways. Nutrient pollution, caused by excess nitrogen and phosphorus in water or air, is the number-one threat to water quality worldwide and can cause algae blooms, a toxic soup of blue-green algae that can be harmful to people and wildlife.

Sewage and Wastewater

Used water is wastewater. It comes from our sinks, showers, and toilets (think sewage) and from commercial, industrial, and agricultural activities (think metals, solvents, and toxic sludge). The term also includes stormwater run-off which occurs when rainfall carries road salts, oil, grease, chemicals, and debris from impermeable surfaces into our waterways.

More than 80 percent of the world's wastewater flows back into the environment without being treated or reused, according to the United Nations; in some least-developed countries, the figure tops 95 percent. In the United States, wastewater treatment facilities process about 34 billion gallons of wastewater per day. These facilities reduce the amount of pollutants such as pathogens, phosphorus, and nitrogen in sewage, as well as heavy metals and toxic chemicals in industrial waste, before discharging the treated waters back into waterways. That's when all goes well. But, our nation's aging and easily overwhelmed sewage treatment systems also release more than 850 billion gallons of untreated wastewater each year.

Oil Pollution

Big spills may dominate headlines, but consumers account for the vast majority of oil pollution in our seas, including oil and gasoline that drips from millions of cars and trucks every day. Moreover, nearly half of the estimated 1 million tons of oil that makes its way into marine environments each year comes not from tanker spills but from land-based sources such as factories, farms, and cities. At sea, tanker spills account for about 10 percent of the oil in waters around the world, while regular operations of the shipping industry contribute about one-third. Oil is also naturally released from under the ocean floor through fractures known as seeps.

Radioactive Substances

Radioactive waste is any pollution that emits radiation beyond what is naturally released by the environment. It's generated by uranium mining, nuclear power plants, and the production and testing of military weapons, as well as by universities and hospitals that use radioactive materials for research and medicine. Radioactive waste can persist in the environment for thousands of years, making disposal a major challenge. Consider the decommissioned Hanford nuclear weapons production site in Washington where clean-up of million gallons of radioactive waste is expected to cost more than $100 billion and last through 2060.

WHAT ARE THE EFFECTS OF WATER POLLUTION?

Water pollution kills. In fact, it caused 1.8 million deaths in 2015, according to a study published in the medical journal, *The Lancet*. Contaminated water can also make you ill. Every year, unsafe water sickens about 1 billion people. And low-income communities are disproportionately at risk because their homes are often closest to the most polluting industries.

Waterborne pathogens, in the form of disease-causing bacteria and viruses from human and animal waste, are a major cause of illness from contaminated drinking water. Diseases spread by unsafe water include cholera, giardia, and typhoid. Even in wealthy nations, accidental or illegal releases from sewage treatment facilities, as well as run-off from farms and urban areas, contribute harmful pathogens to waterways. Thousands of people across the United States are sickencd every year by Legionnaires' disease (a severe form of pneumonia contracted from water sources like cooling towers and piped water), with cases cropping up from California's Disneyland to Manhattan's Upper East Side.

Even swimming can pose a risk. Every year, 3.5 million Americans contract health issues such as skin rashes, pinkeye, respiratory infections, and hepatitis from sewage-laden coastal waters, according to EPA estimates.

ON THE ENVIRONMENT

In order to thrive, healthy ecosystems rely on a complex web of animals, plants, bacteria, and fungi—all of which interact, directly or indirectly, with each other. Harm to any of these organisms can create a chain effect, imperiling entire aquatic environments.

When water pollution causes an algal bloom in a lake or marine environment, the proliferation of newly introduced nutrients stimulates plant and algae growth, which in turn reduces oxygen levels in the water. This dearth of oxygen suffocates plants and animals and can create "dead zones," where waters are essentially devoid of life. In certain cases, these harmful algal blooms can also produce neurotoxins that affect wildlife, from whales to sea turtles.

Chemicals and heavy metals from industrial and municipal wastewater contaminate waterways as well. These contaminants are toxic to aquatic life—most often reducing an organism's life span and ability to reproduce—and make their way up the food chain as predator eats prey. That's how tuna and other big fish accumulate high quantities of toxins, such as mercury.

Marine ecosystems are also threatened by marine debris which can strangle, suffocate, and starve animals. Much of this solid debris, such as plastic bags and soda cans, gets swept into sewers and storm drains and eventually out to sea, turning our oceans into trash soup and sometimes consolidating to form floating garbage patches. Discarded fishing gear and other types of debris are responsible for harming more than 200 different species of marine life.

Meanwhile, ocean acidification is making it tougher for shellfish and coral to survive. Though they absorb about a quarter of the carbon pollution created each year by burning fossil fuels, oceans are becoming more acidic. This process makes it harder for shellfish and other species to build shells and may impact the nervous systems of sharks, clownfish, and other marine life."

What Can You Do to Prevent Water Pollution?

It's easy to tsk-tsk the oil company with a leaking tanker, but we're all accountable to some degree for today's water pollution problem. Fortunately, there some simple ways you can prevent water contamination or at least limit your contribution to it:

Reduce your plastic consumption and reuse or recycle plastic when you can.

Properly dispose of chemical cleaners, oils, and non-biodegradable items to keep them from ending up down the drain.

Maintain your car so it doesn't leak oil, antifreeze, or coolant.

If you have a yard, consider landscaping that reduces run-off and avoid applying pesticides and herbicides.

If you have a pup, be sure to pick up its poop.

Use Your Voice

One of the most effective ways to stand up for our waters is to speak out in support of the Clean Water Rule, which clarifies the Clean Water Act's scope and protects the drinking water of one in three Americans.

Tell the federal government, the U.S. Army Corps of Engineers, and your local elected officials that you support the Clean Water Rule. Also, learn how you and those around you can get involved in the policymaking process. Our public waterways serve every American. We should all have a say in how they're protected.

Endnotes

Foreword

1 Jagai, J. S., Li, Q., Wang, S., Messier, K. P., Wade, T. J., and Hilborn, E. D., *Environmental Health Perspectives*, September 2015, 873-879: "Extreme Precipitation and Emergency Room Visits for Gastrointestinal Illness in Areas with and without Combined Sewer Systems: An Analysis of Massachusetts Data, 2003–2007."

   "The rate of ER visits for GI illness was associated with extreme precipitation in the area with CSO discharges to a drinking water source. Our findings suggest an increased risk for GI illness among consumers whose drinking water source may be impacted by CSOs after extreme precipitation."

Chapter 4

1 The lower Merrimack from Haverhill to Newburyport was undoubtedly the greatest shipbuilding center of New England. Close to 1115 vessels were constructed and registered there between 1793 and 1815. From Morison, S. E., *Maritime History of Massachusetts* (1921, Houghton Mifflin, reprinted in 1979 by Northeastern University Press), p. 101.

2 Currier, J. J., *The History of Newburyport Mass., 1764–1905, Vols. 1–2* (Reprinted edition, Somersworth; New Hampshire Publishing Co., 1977), p. 22.

3 Doyle, J. F., *Life in Newburyport, 1950–1985: A Collection of News Events, City Affairs, and Memories from the Last Half of the Twentieth Century* (Portsmouth, N.H.: Peter E. Randall, 2010), p. 236.

Chapter 6

1 Accessed from bartleby.com/270/13/154.html

Chapter 7

1 Thoreau, H. D., excerpted from an essay published by the National Park Service, U.S. Department of the Interior, "Experience Your America," November 9, 2018.

CHAPTER 8

1   Selected sources on Wikipedia site dedicated to Haverhill, en.wikipedia.org/wiki/Haverhill,_ Massachusetts

CHAPTER 9

1   Selected sources on Wikipedia site dedicated to Lawrenceen.wikipedia.org/wiki/Lawrence,_ Massachusetts

CHAPTER 11

1   Selected sources on Wikipedia site dedicated to Lowell, en.wikipedia.org/wiki/Lowell,_ Massachusetts

CHAPTER 12

1   Selected sources on Wikipedia site dedicated to Manchester, N.H.https://en.wikipedia.org/ wiki/Manchester,_New_Hampshire

BIBLIOGRAPHY

Note: I have a weekly podcast that airs on local radio and TV. It is titled *Life Along the Merrimack*. I have hosted more than seventy sessions with people involved with the river. Some of my "general knowledge" and quotes come from these conversations.

American Rivers, "Merrimack River (MA, NH)" (written assessment concerning ecological threats Washington, D.C., 2016)

Atkinson, J., *Massacre on the Merrimack: Hannah Dustin's Captivity and Revenge in Colonial America* (Guilford, Conn.: Lyons Press, 2015)

Cheney, R., *Maritime History of Merrimac Shipbuilding* (Newburyport, Mass.: Newburyport Press, 1964)

Cheney, R., *The Cheney Collection* (Newburyport, Mass.: Custom House Maritime Museum, 1992)

Chernow, R., *Alexander Hamilton* (New York: Penguin Books, 2004)

Cole, D. B., *Immigrant City, Lawrence, Mass., 1845–1921* (Chapel Hill, North Carolina: University of North Carolina Press, 2002)

Cowley, C., *A History of Lowell, Massachusetts; Lowell, Mass.* (Jazzybee Verlag: 2017)

Currier, J. J., *History of Newbury, Mass, 1635–1902* (Boston: Damrell & Upham, 1902)

Daily News, Newburyport (Archives of this newspaper were consulted regularly, 2019–2020)

Dengler, E., Khalife, K., and Skulski, K., *Lawrence, Massachusetts* (Charleston, S.C.: Arcadia Publishing, 1995)

DiZoglio, state Sen. D., podcast interview, *Life Along the Merrimack* methods to halt pollution (2019)

DiZoglio, state Sen. D., phone interview, methods to improve river (2020)

Doyle, J. F., *Life in Newburyport, 1950–1985* (Portsmouth: Peter Randall Publisher, 2010)

Eagle-Tribune, North Andover, Mass. (Archives of this newspaper were consulted regularly, 2019–2020)

Eaton, A., *The Amoskeag Manufacturing Company: A History of Enterprise on the Merrimack River* (Charleston, S.C.: Arcadia Publishing Co., 2015)

Eigerman, J., city councilor, podcast interview, Newburyport, *Life Along the Merrimack* (2019)

Evans-Dale, L. and Utterbach, M. (eds.), *The ABCs of Newburyport Maritime History* (Newburyport, Mass., booklet for Custom House Maritime Museum, 2014)

Fiorentini, J. J., mayor of Haverhill, phone interview, *The Use of Merrimack River in the Rebuilding of Haverhill* (2020)

Graovac, D., chair of the board of the Merrimack River Watershed Council, podcast interview, *Life Along the Merrimack* (2019)

Haggerty, R., "Reinventing a River," *American Heritage* (Rockville, M.D.: 2003)

Hendrickson, D., *Nautical Newburyport: A History of Captains, Clipper Ships and the Coast Guard* (Charleston, S.C.: The History Press, 2017)

Janson, R., former spokesperson for the Environmental Protection Agency in New England, phone interview (subject was the value of Sen. George Mitchell to the 1987 Amendments of Clean Water Act, 2020)

Hudon, P., *Lower Merrimack: The Valley and Its Peoples; An Illustrated History* (Sun Valley, Calif.: American Historical Press, 2004)

Hughes, B., *The Artwork of Blake Hughes* (Newburyport, the Newburyport Printmaker, 2011)

Kelcourse, state Rep. J., phone interview, political action to improve river quality (2020)

Macone, J., outreach historian with Merrimack River Watershed Council, podcast interviews, *Life Along the Merrimack* (2019–2020)

Maine Historical Society, Maine Memory Network, "Clean Water: Muskie and the Environment," essay (Portland, 2005)

Meader, J. W., *The Merrimack River: Its Sources and Its Tributaries* (Boston: B.B. Russell, 1869)

Morison, S. E., *The Maritime History of Massachusetts, 1783–1860* (Boston, reprint, Northeastern University Press, 1979)

Mitchell, Sen. G., speaking at the Commemoration of the thirtieth Anniversary of the Clean Water Act (Committee on Environment and Public Works, U.S. Senate, 107th Congress, Second Session, Washington, D.C., 2002)

Prendergast, J., *Lowell: Images of America* (Charleston, S.C.: Arcadia Publishing, 1996)

Roberts, J., *The Connecticut River from the Air* (Guilford, Conn.: Globe/Pequot Publishing, 2018)

Samson, G., *Manchester: The Mills and The Immigrant Experience* (Portsmouth, N.H.: Arcadia Publishing, 2000)

Sciolino, E., *The Seine: The River That Made Paris* (New York: W.W. Norton & Company, Inc., 2020)

Shumway, C., outgoing executive director of the Merrimack River Watershed Council, "We Need Everyone's Help on the Merrimack River Now" (July 25, 2016)

Sparks, D., "The Course of the Merrimack, My Time Among the Voyagers," *Merrimack Valley Magazine* (Methuen, Mass.: 2019)

Theokas, A. C., *Lowell Through Time* (Charleston, S.C.: Arcadia Publishing, 2018)

Thoreau, H. D., *A Week on the Concord and Merrimack Rivers, 1849* (various reprinted editions)

Turner, C. W., *Remembering Haverhill: Stories from the Merrimack Valley* (Charleston, S.C.: American Chronicles, 2008)

Union Leader, Manchester, N.H. (Archives of this newspaper were consulted regularly)

Woodworth, G., Newburyport historian, podcast interview, *Life Along the Merrimack* (2020)

About the Author

DYKE HENDRICKSON is an author-journalist living in Newburyport, Mass., birthplace of the Coast Guard. *Merrimack, the Resilient River* is his sixth book.

Several years ago, he wrote a multi-part series for *The Daily News*, the local newspaper in Newburyport, on the 250th history of the city. Hendrickson researched and wrote *Nautical Newburyport: A Story of Captains, Clipper Ships and the Coast Guard*, published by The History Press in 2017. He then wrote *New England Coast Guard Stories: Remarkable Mariners*, which was published in March 2020, also by the History Press. *Merrimack, the Resilient River* will mark the third in his Merrimack trilogy.

He is currently a historian with the Merrimack River Watershed Council. In that role, he speaks on Zoom to

Author Dyke Hendrickson.

clubs, associations, and historical gatherings on the history of the Coast Guard and the Merrimack River. He also hosts a weekly podcast titled *Life Along the Merrimack*, during which he interviews local and state officials about the health and history of the river. The half-hour show goes out live on local radio and cable TV; it is archived on YouTube.

He lives not from the sea with his wife, Vicki. They have two children, Leslie, of New York, and Drew, of Somerville, Mass.; a grandson, Nico; and a daughter-in-law, Natalia.

The author is a graduate of Franklin and Marshall College with a degree in history, and he did graduate work at the University of Maine, Orono. He has been a writer and editor with *The Portland Press Herald*, *The New Orleans Times-Picayune*, and *The Boston Herald*. Other publications he has written for include *USA Today*, *The Boston Globe*, and *Tennis* magazine. Most recently, he was the waterfront reporter for *The Daily News* in Newburyport.